Mohamed Elzagheid
Biomacromolecules

Also of Interest

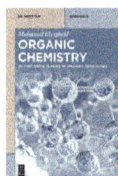

Organic Chemistry.
25 Must-Know Classes of Organic Compounds
Mohamed Elzagheid, 2024
ISBN 978-3-11-138199-2, e-ISBN 978-3-11-138275-3

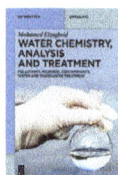

Water Chemistry, Analysis and Treatment.
Pollutants, Microbial Contaminants, Water and Wastewater Treatment
Mohamed Elzagheid, 2024
ISBN 978-3-11-133242-0, e-ISBN 978-3-11-133246-8

Chemical Technicians.
Good Laboratory Practice and Laboratory Information Management
Systems
Mohamed Elzagheid, 2023
ISBN 978-3-11-119110-2, e-ISBN 978-3-11-119149-2

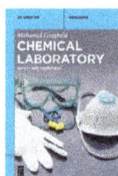

Chemical Laboratory.
Safety and Techniques
Mohamed Elzagheid, 2022
ISBN 978-3-11-077911-0, e-ISBN 978-3-11-077912-7

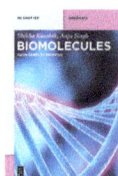

Biomolecules.
From Genes to Proteins
Shikha Kaushik, Anju Singh, 2023
ISBN 978-3-11-079375-8, e-ISBN 978-3-11-079376-5

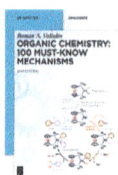

Organic Chemistry: 100 Must-Know Mechanisms
Roman Valiulin, 2023
ISBN 978-3-11-078682-8, e-ISBN 978-3-11-078683-5

Mohamed Elzagheid

Biomacromolecules

Carbohydrates, Lipids, Proteins and Nucleic Acids

2nd Edition

DE GRUYTER

Author
Prof. Dr. Mohamed Elzagheid
Royal Commission for Jubail and Yanbu
Jubail Industrial College
Jubail Industrial City
Saudi Arabia
and
Center for Research and Strategic Studies
Libyan Authority for Scientific Research
Tripoli, Libya
melzagheid@gmail.com

ISBN 978-3-11-158298-6
e-ISBN (PDF) 978-3-11-158327-3
e-ISBN (EPUB) 978-3-11-158397-6

Library of Congress Control Number: 2024945563

Bibliographic information published by the Deutsche Nationalbibliothek
The Deutsche Nationalbibliothek lists this publication in the Deutsche Nationalbibliografie;
detailed bibliographic data are available on the Internet at http://dnb.dnb.de.

© 2025 Walter de Gruyter GmbH, Berlin/Boston
Cover image: Christoph Burgstedt/iStock/Getty Images Plus
Typesetting: Integra Software Services Pvt. Ltd.

www.degruyter.com
Questions about General Product Safety Regulation:
productsafety@degruyterbrill.com

This book is dedicated to everyone who has supported me in my quest to write and share my expertise.

Preface

The second edition of the macromolecular chemistry textbook broadens in two areas. The current book delves deeper into biomacromolecules, while another covers broader issues in man-made polymers.

Biomacromolecules include carbohydrates, lipids, proteins, and nucleic acids. The carbohydrates chapter looks at the structural formulas and cyclic forms of monosaccharides, as well as their synthesis and breakdown. Cyclization, enolization, isomerization, tautomerization, mutarotation, and epimerization are all briefly described. The second chapter covers triglycerides, steroids, vitamins, and their constituents. On the other hand, the third chapter examines the primary structure of proteins, including amino acid properties, peptide bond formation, and peptide synthesis. It also addresses secondary, tertiary, and quaternary structures. The book concludes with a chapter on nucleic acids, which covers the chemistry of nucleosides and oligonucleotides as well as very interesting topics like genetic code, DNA secret code, Polymerase Chain Reaction (PCR), and DNA fingerprinting.

The aforementioned topics are covered at the chemistry student level and presented in an easily understood manner, making them understandable to other students enrolled in comparable or different programs.

This book is suitable for both graduate and advanced undergraduate students, as well as researchers, chemists, and lab technicians working in organic, bioorganic, biochemistry, and biological chemistry laboratories. The book's language is simple, concise, and understandable to all readers, including those with only a basic background of macromolecular chemistry.

I hope teachers find this book useful as a reference for their classes, and that students appreciate learning about the topics in the various sections.

Mohamed Ibrahim Elzagheid, Chemistry Professor
Waterloo, Ontario, Canada
2024

https://doi.org/10.1515/9783111583273-202

Acknowledgment

First and first, I'd like to convey my heartfelt gratitude to my entire family for their unwavering help and encouragement throughout my academic career, whether it was teaching, conducting research, or writing.

I also want to thank my colleagues from the chemical engineering department and the Libyan authority for scientific research for their vital support. Special thanks to Professor Adam Elzagheid, the managing director of the Libyan Biotechnology Research Centre, for his help and backing.

Finally, a sincere thanks to the entire publishing team, especially Ute Skambraks, Helene Chavaroche, and Chandhini Magesh, for their invaluable support.

https://doi.org/10.1515/9783111583273-203

The Author

Mohamed Elzagheid is a chemistry associate professor at the Royal Commission for Jubail and Yanbu, as well as a professor and consultant with the Libyan Authority for Scientific Research.

During his 30-year career at Turku University in Finland, McGill University, SynPrep Inc. in Montreal, Canada, and Jubail Industrial College in Saudi Arabia, he was directly and indirectly involved in the education of laboratory technicians and chemists, as well as supervising numerous undergraduate and graduate chemistry students.

He has made important contributions to numerous short-term and long-term training programs for Saudi corporations, as well as teaching a wide range of university courses at different levels. The courses provided include basic and advanced Organic Chemistry, Polymer Chemistry, Introduction to Macromolecule Chemistry, Biochemistry, Laboratory Techniques, Safety in Chemical Laboratories, Technician Responsibility, and Water and Wastewater Treatment.

He also chaired the Research, Projects, Publications, and Academic Promotion Team, Academic Promotion Committee, Curriculum Development Committee, Industrial Chemistry Technology Program Advisory and Evaluation Committee, CTAB Steering Accreditation Committee, Industrial Outreach Committee, and Chemical Engineering Department Safety Committee at Jubail Industrial College.

Dr. Elzagheid is the author of seven textbooks: **Introductory Organic Chemistry**, **Thoughts on Organic Chemistry**, **Macromolecular Chemistry**: Natural and Synthetic Polymers, **Chemical Laboratory Safety, Techniques**, and **Chemical Technicians**: Good Laboratory Practice and Laboratory Information Management Systems, **Water Chemistry, Analysis and Treatment**: Pollutants, Microbial Contaminants, Water and Wastewater Treatment, and **Organic Chemistry**: 25 Must-Know Classes of Organic Compounds.

His work at Turku University in Finland, McGill University in Canada, and JIC in the Kingdom of Saudi Arabia has helped him develop a solid reputation in chemistry and chemical education, as demonstrated by his research papers and publications.

https://doi.org/10.1515/9783111583273-204

Contents

Chapter 1
Introduction

1.1 Macromolecules Versus Polymers

A macromolecule is a big polymeric or nonpolymeric molecule with a high molecular mass. Polymers are large molecules with a high molecular weight made up of small monomeric repeating units (the repeat units like wagons in a train) that are typically bonded or connected by *O*-glycosidic bond, which is found in carbohydrates, *N*-glycosidic bond, which is found in nucleic acids, and peptide bond, which is found in polypeptides and proteins. Structures of these bonds are presented in Figure 1.1.

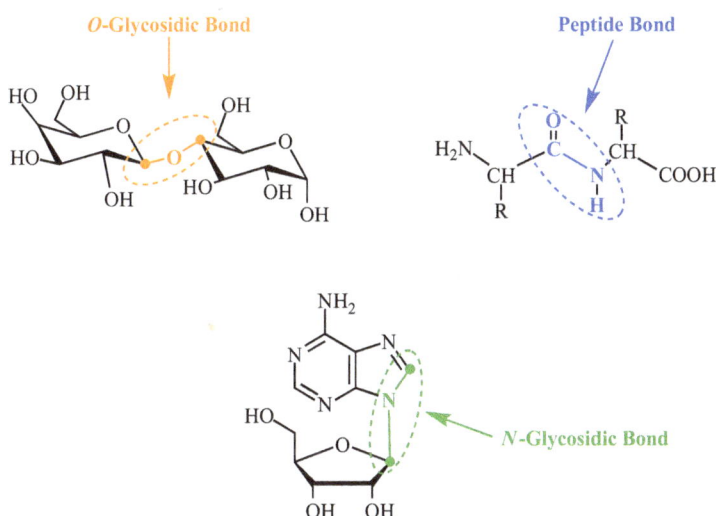

Figure 1.1: Examples of *O*-glycosidic, *N*-glycosidic, and peptide (linkage) bonds.

Carbohydrates are macromolecules that are categorized as polymers since they are composed of repeated monosaccharides. Similarly, protein is a polymer due to its repeating amino acid sequence. However, because a fat (lipid) is composed of a distinct set of molecules that are not repeating monomers, it cannot be described as a polymer. Its composition consists of three fatty acids and one molecule of glycerol. The fatty acid types differ, yet there is just one glycerol. Figures 1.2 and 1.3 show examples of macromolecules.

Maitotoxin, a macromolecule, is a strong marine toxin associated with ciguatera poisoning found in tropical and subtropical Pacific Ocean locations. It is one of the largest natural nonpolymeric substances. Figure 1.4 illustrates yet another macromolecule that cannot be categorized as a polymer.

https://doi.org/10.1515/9783111583273-001

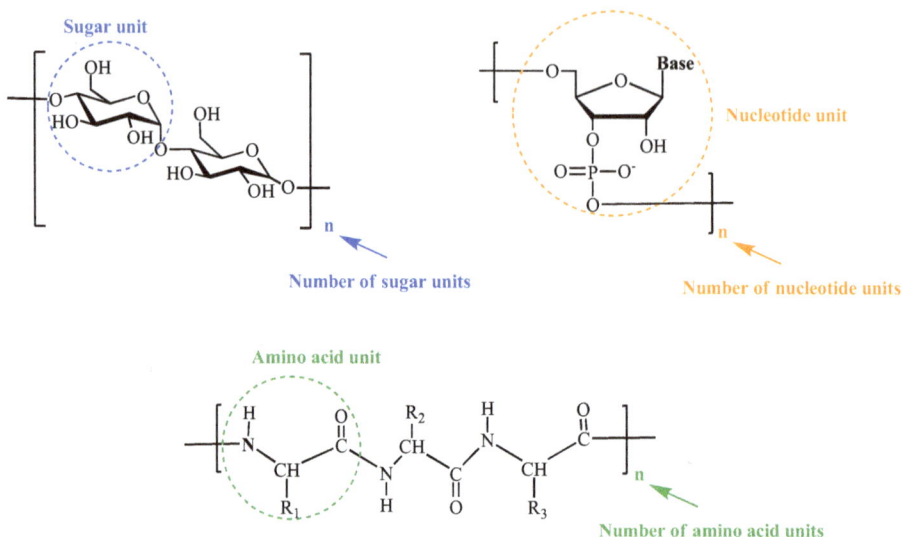

Figure 1.2: Examples of macromolecules that can be classified as polymers.

In general, the use of the term "macromolecule" covers biomacromolecules (biopolymers) such as proteins, polysaccharides, nucleic acids, and lipids, and synthetic polymers such as polyethylene oxide, polyethylene, and polypropylene. Biomacromolecules are natural polymers that are found in nature, and some of their analogs can be made in the laboratory. Synthetic polymers are man-made polymers that are synthesized from small organic and inorganic molecules, and most of them are made from oil and gas products.

1.2 Biomacromolecules (Biopolymers or Natural Polymers)

Examples of biomacromolecules are carbohydrates (sugars), proteins (peptides and polypeptides), and nucleic acids (DNA and RNA), and the monomer units are monosaccharides, amino acids, and nucleotides, respectively. Lipids (fats and oils) have fatty acids and glycerol as building blocks (Table 1.1).

Carbohydrates can either have simple structures as in monosaccharides or complex structures as in disaccharides, oligosaccharides, and polysaccharides (Figure 1.5). Their size can range from four carbons as in tetroses to six carbons as in hexoses. They can also exist as aldoses or ketoses based on the functional groups they have in addition to polyhydroxy groups.

Lipids are a diverse group of organic compounds. They are insoluble in water but dissolve in nonpolar solvents such as ether, hexane, and acetone. Most lipids have fatty acids in their structure, except steroids, which are made of tetracyclic carbon rings. Fatty

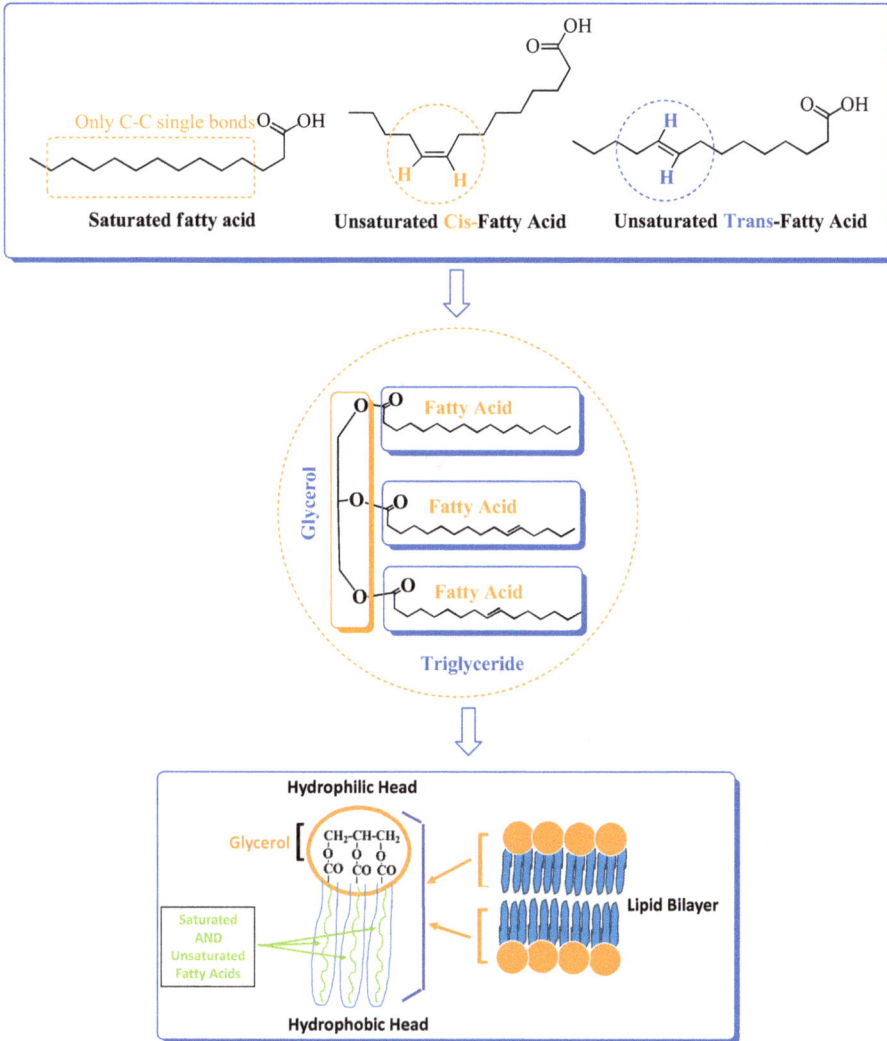

Figure 1.3: An illustration of a lipid macromolecule's simplified structure.

acids are long-chain carboxylic acids, and they can be saturated or unsaturated based on the presence or absence of double bonds in the structure. Triglycerides are formed when glycerol is esterified by three fatty acids. Lipids can be classified into waxes, triglycerides, phospholipids, sphingolipids, eicosanoids, and steroids. Their levels of structure are presented in Figure 1.6.

Proteins are the polymers of amino acids. In the structure, each amino acid has two functional groups, namely an amine and a carboxyl group. Based on the side chain, amino acids can have different names and different physical and chemical

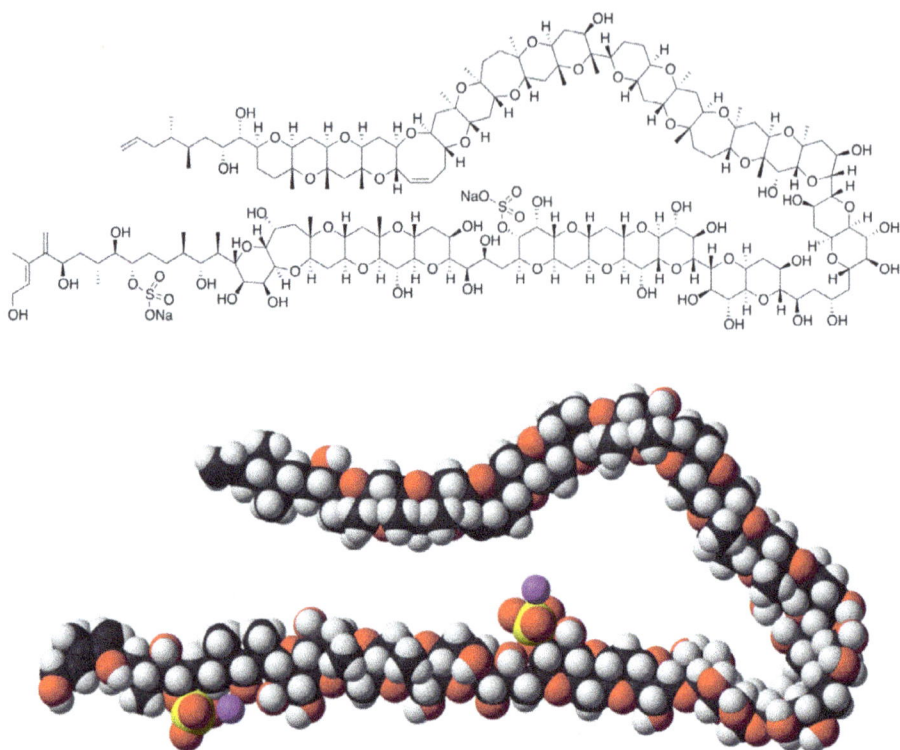

Figure 1.4: Maitotoxin macromolecule structure (image credit: https://en.wikipedia.org/wiki/Maitotoxin).

Table 1.1: Examples of biomacromolecules and their monomeric units.

Biomacromolecules	Building blocks	Atoms present in the structure	Examples
Carbohydrates (sugars or polysaccharides)	Monosaccharides	C, H, O	Glucose, sucrose, starch, cellulose, and chitin
Proteins (polypeptides)	Amino acids	C, H, O, N	Keratin, wool, collagen, and silk
Nucleic acids	Nucleotides	C, H, O, N, P	DNA and RNA
Lipids (fats and oils)	Fatty acids and glycerol	C, H, O	Waxes and steroids
	Fatty acids, phosphoric acid, and glycerol	C, H, O, P	Phospholipids

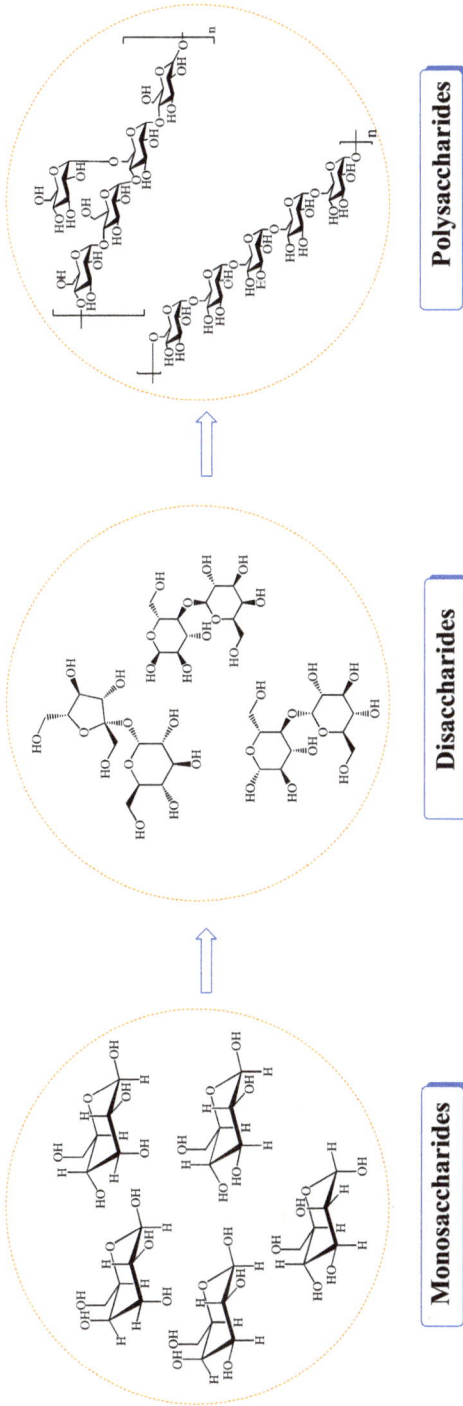

Figure 1.5: Levels of carbohydrate structure.

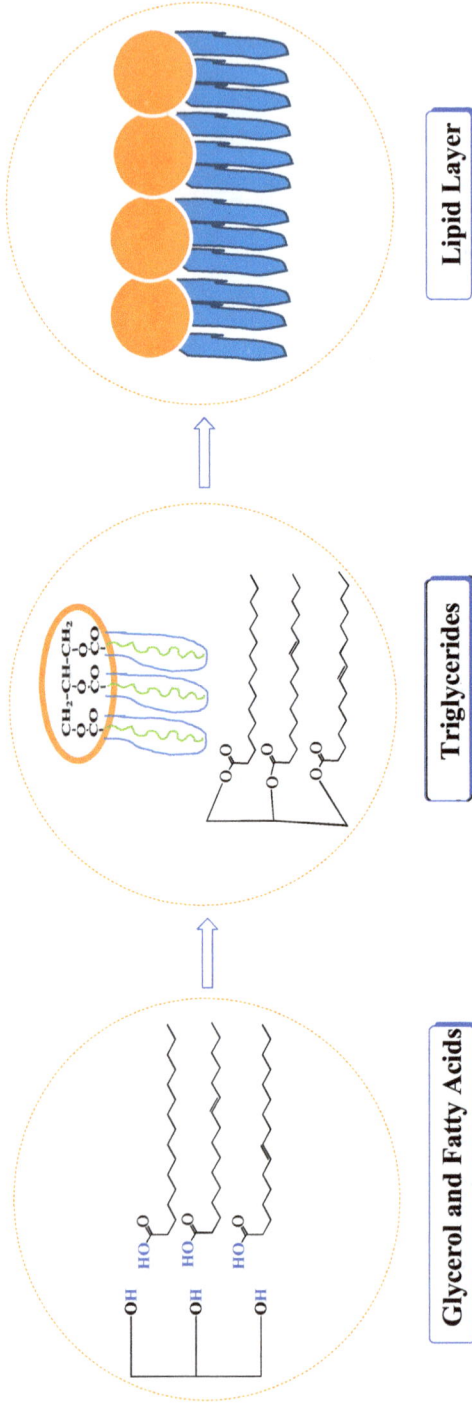

Figure 1.6: Levels of lipid structure.

properties. Amino acids can be polar, nonpolar, acidic, or basic. The names of amino acids can be either abbreviated by a three-letter code or a one-letter code. For example, the name "alanine" can be abbreviated as Ala or A.

All amino acids are chiral except glycine because it has only three different groups attached to the central carbon. They can form dipeptides by the condensation of two amino acids and polypeptides by the condensation of more amino acids. Polypeptides are usually made in the laboratory through solid-phase peptide synthesis. The levels of protein structure are presented in Figure 1.7.

Nucleic acids are the polymers of nucleotides. There are two types of nucleic acids, namely ribonucleic acid (RNA) and deoxyribonucleic acid (DNA). Each nucleotide is made of three parts: nucleobase, pentose sugar, and phosphate group. There are five types of nucleobases: uracil, thymine, cytosine, guanine, and adenine. Pentose sugars are deoxyribose and ribose. There are different types of modified nucleic acids. Among those are the peptide (or peptido) nucleic acid (PNA), morpholino nucleic acid (MNA) glycol/glycerol nucleic acids, threose nucleic acid, and 4'-thionucleic acids (4'-SDNA). Their levels of structure are shown in Figure 1.8.

1.3 Chemical Bonding

The two main ways in which atoms can be combined to form molecules are electrovalent bonding (forming ionic bonds) or covalent bonding (forming covalent bonds). Some molecules contain both electrovalent and covalent bonds, but many have just one or the other type. There are also much weaker attractions between atoms in molecules called hydrogen bonds. Hydrogen bonding is usually found in nucleic acids and proteins.

1.3.1 Bonds Classification

Bonds are classified on the basis of their electronegativities into nonpolar, covalent, ionic, and polar covalent. If two atoms have an electronegativity difference of 0.4 or less, they form a nonpolar covalent bond. If two atoms have an electronegativity difference between 0.4 and 1.8, they form a polar covalent bond. If two atoms have an electronegativity difference of more than 1.8, they form an ionic bond as given in Table 1.2.

1.3.2 Electrovalent (Ionic) Bond

Electrovalent (ionic) bonds are formed by an electrical attraction between positively charged cations and negatively charged anions, as shown in Figure 1.9.

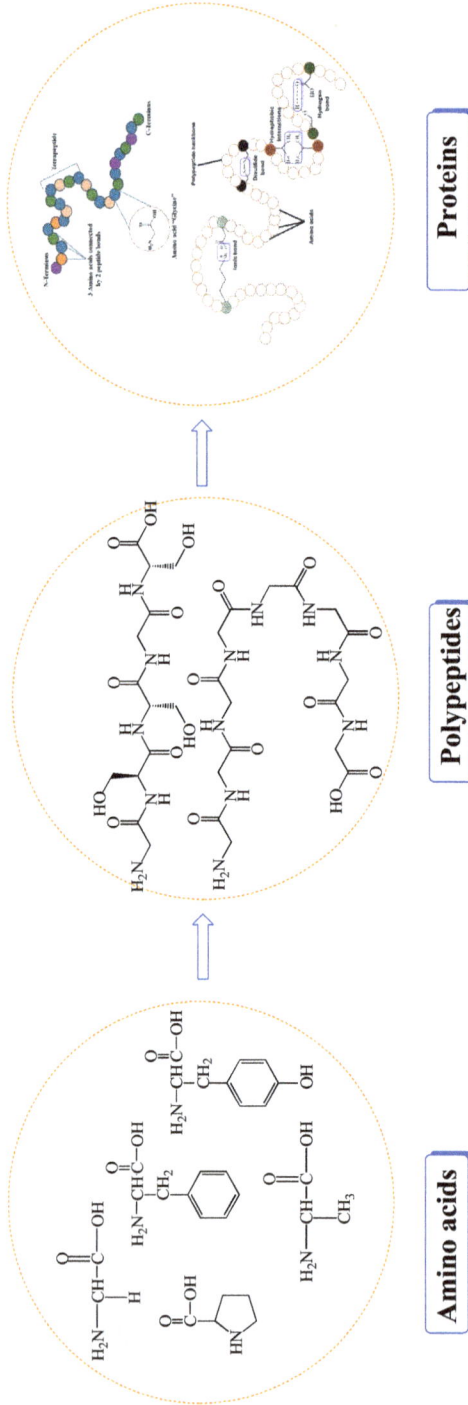

Figure 1.7: Levels of protein structure.

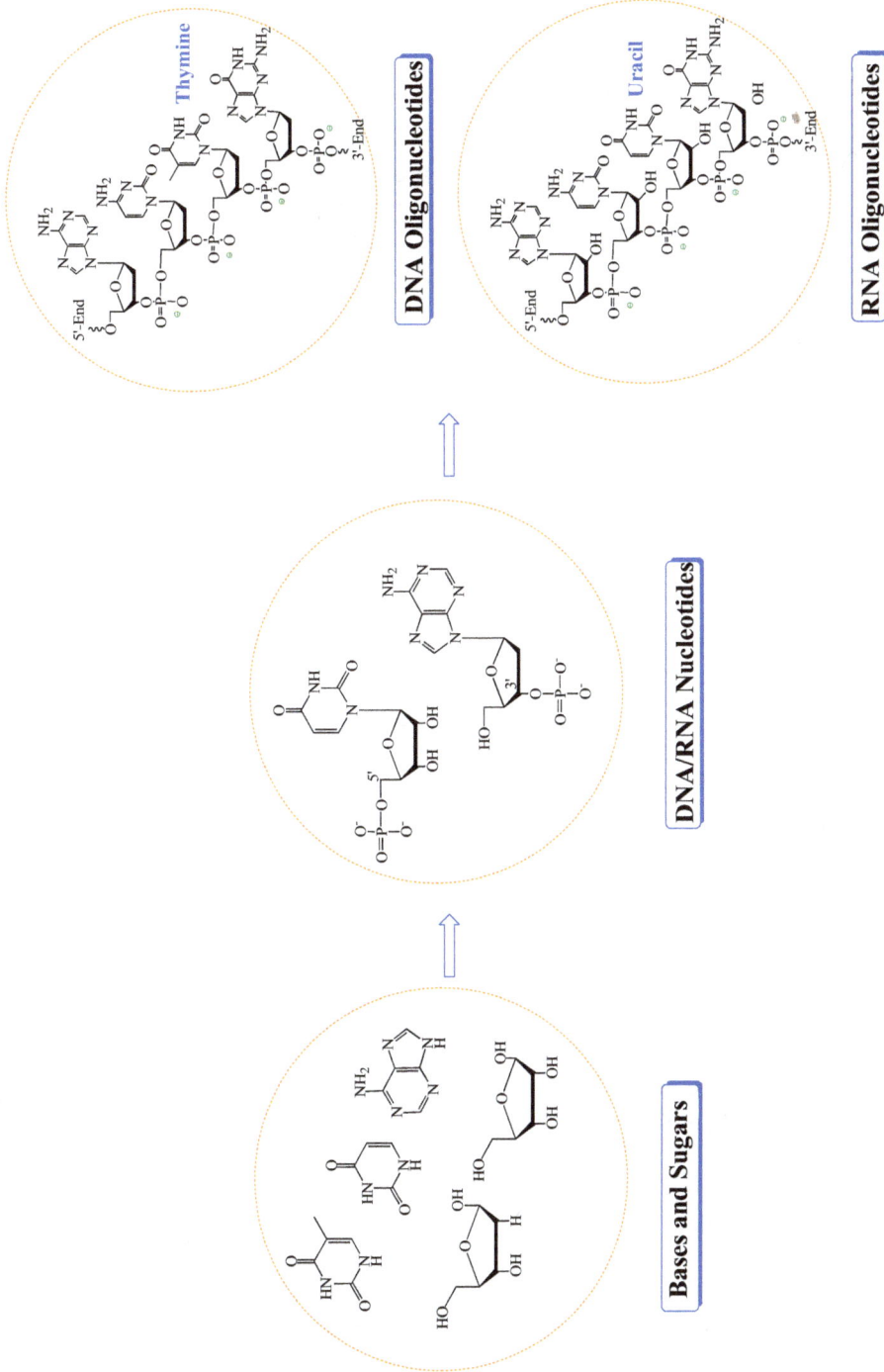

Figure 1.8: Levels of nucleic acid structure.

Table 1.2: Classification of bonds.

Bond type	Difference in electronegativity	Examples
Nonpolar covalent	<0.4	Cl–Cl (3.16 – 3.16 = 0) H–H (2.20 – 2.20 = 0) O=O (3.44 – 3.44 = 0) H–S–H (2.58 – 2.20 = 0.38)
Polar covalent	0.4–1.8	H–O–H (3.44 – 2.20 = 1.24) H–F (3.98 – 2.20 = 1.78) O=C=O (3.44 – 2.55 = 0.89) NH_3 (3.04 – 2.20 = 0.84)
Ionic	>1.8	NaCl (3.16 – 0.93 = 2.23) KBr (2.96 – 0.82 = 2.14) KF (3.98 – 0.82 = 3.16) KI (2.66 – 0.82 = 1.84)

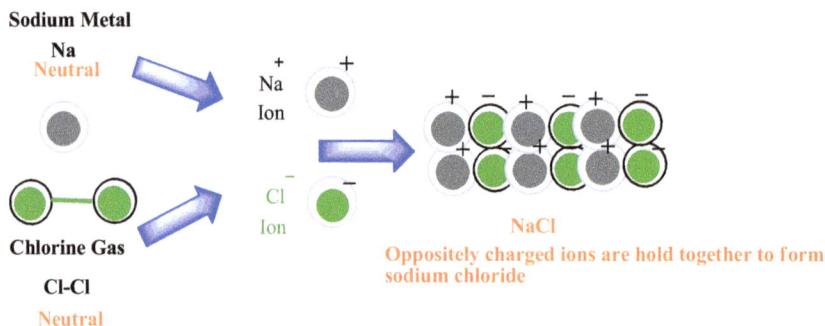

Figure 1.9: NaCl ionic bond formation.

1.3.3 Covalent Bond

Covalent bonds are formed by sharing electrons between atoms. An example is the methane gas, CH_4, as shown in Figure 1.10.

1.3.4 Hydrogen Bond

Hydrogen bonds are formed by attractions between the positively charged hydrogen atoms with either oxygen, nitrogen, fluorine, or, in some cases, sulfur atoms, as shown in Figure 1.11.

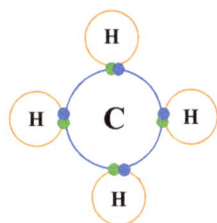

- ● Electron from carbon
- ● Electron from hydrogen

Figure 1.10: Covalent bond electron sharing in methane.

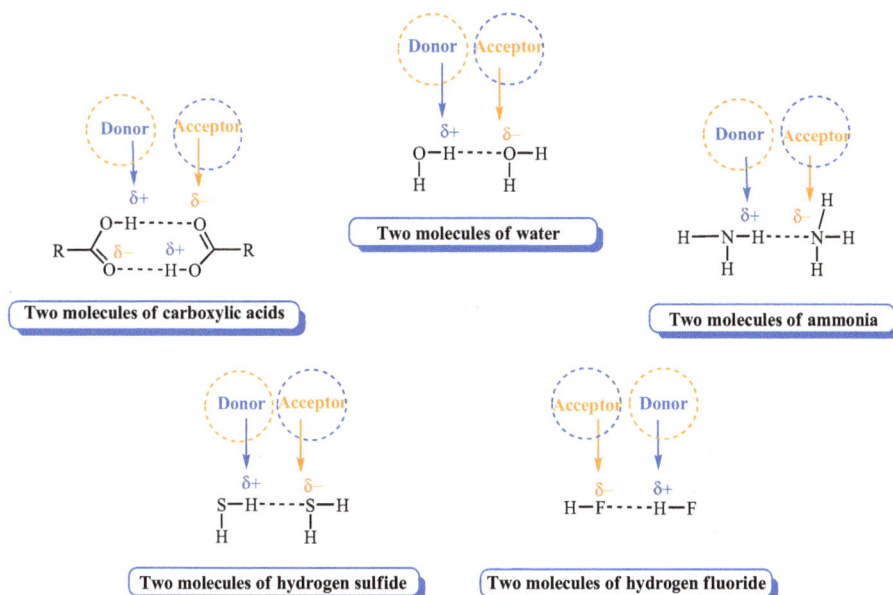

Figure 1.11: Hydrogen bond formation between different molecules.

1.4 Types of Chemical Reactions in Biomacromolecules

1.4.1 Glycoside Bond Formation (Glycosidation Reaction)

Glycosidation is simply the addition of sugar molecules to noncarbohydrate moieties like alcohols. Fischer glycosidation is an excellent example of how an aldose or ketose interacts with an alcohol in the presence of an acid catalyst to form a glycoside:

Sugar —OH $\xrightarrow[\text{H}^+]{\text{ROH}}$ Glycoside —OR

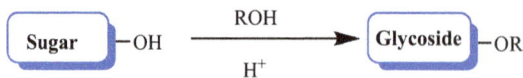

1.4.2 Triglyceride Formation (Esterification Reaction)

Triglyceride formation is an esterification process in which the hydroxyl (–OH) groups of the glycerol and carboxylic (–COOH) groups of the three long-chain carboxylic acids (fatty acids) join to produce a triglyceride. In other words, the triesters in triglycerides are formed by the condensation of glycerol and three fatty acids, as well as the dehydration of water molecules:

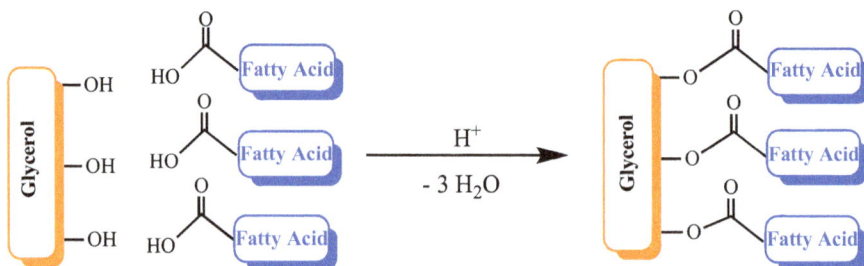

1.4.3 Peptide Bond Formation (Condensation Reaction)

Peptide bonds are generated when two amino acids unite to form a peptide, polypeptide, or protein. These bonds are created through a process known as dehydration (also known as polymerization), a process by which monomers (amino acids) are bonded to form polymers. As a result, a water molecule is produced.

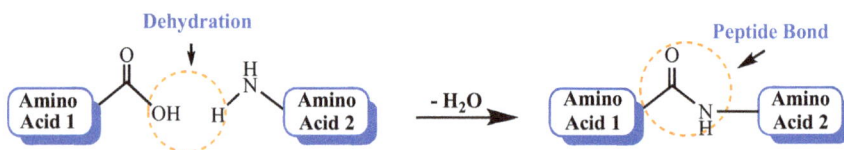

1.4.4 Dinucleotide or Oligonucleotide Formation (Coupling Reaction)

This type of reaction is also called dehydration synthesis, which means "to assemble while losing water." When oligonucleotides are formed, either the sugar's C-5′ phosphate group reacts with the free OH group on the 3′-C or the sugar's C-3′ phosphate group reacts with the free OH group on the 5′-C, resulting in the loss of one water molecule and the formation of the phosphodiester bond. In other words, the phosphate group in one nucleotide creates an ester bond with the OH group on the third or the fifth carbon atom of a sugar unit in another nucleotide or the first nucleoside attached to the solid support.

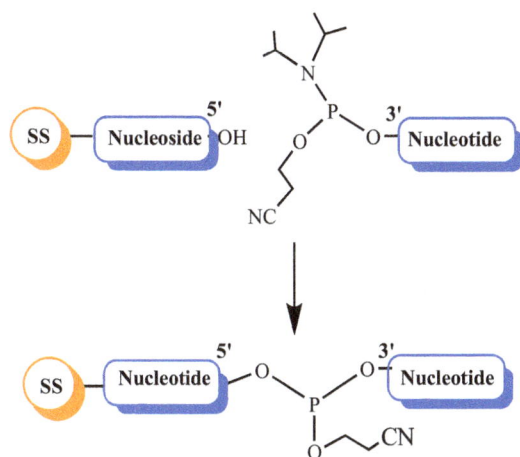

1.5 Drawing Chemical Structures of Biomacromolecules

To draw chemical structures in general by hand or with ChemDraw, utilize the simple and time-saving skeletal formula. The Kekulé formula is more time-consuming due to the huge number of hydrogens and other atoms in the structure. For example, although vitamin A and cholesterol are small molecules, sketching their chemical structures with the Kekulé formula takes some time, whereas the skeletal formula reveals which one is the best fit and takes less time (Figure 1.12).

Drawing the complete representative chemical structures of biomacromolecules is difficult since they are made up of large different building pieces, each of which contains multiple functional groups and, in some cases, a few structural units. So, the simplest technique to achieve a complete structure is to create separate single components and then join them via links. ChemDraw software is one of the greatest tools for creating chemical structures since it allows you to simply create various simplified structures. Examples of selected chemical structures are shown in Figure 1.13.

1.6 Functional Groups in Biomacromolecules

The functional groups in macromolecules are hydroxyl, carbonyl, carboxyl, amino, phosphate (phosphoryl), and sulfhydryl. These groups contribute significantly to the formation of molecules such as DNA, proteins, carbohydrates, and lipids. Carbohydrates typically contain both carbonyl and hydroxyl functional groups. Lipid structures vary, but the most frequent functional groups are ester groups such as carboxylate and phos-

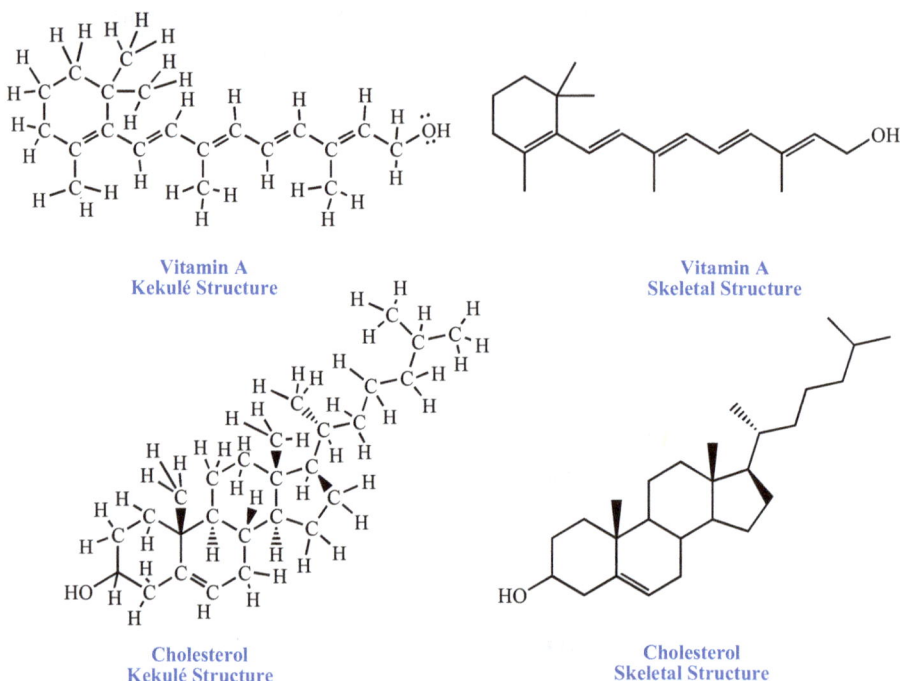

Figure 1.12: Kekulé versus skeletal chemical structures of selected molecules.

phate, in addition to a hydroxyl group. Functional groups found in amino acids include carboxyl, sulfhydryl, and amino, while nucleic acids have phosphate or phosphoryl groups. Figure 1.14 depicts a selection of these functional groups.

1.7 Essential Keywords

Biomacromolecule or biopolymer A large molecule that is essential to biological processes, such as a protein or nucleic acid. It consists of thousands of covalently linked atoms. Many macromolecules are composed of simpler molecules known as monomers.

Carbohydrates Polymers of monosaccharides can form complex structures of disaccharides, oligosaccharides, and polysaccharides.

Chemical bonding Combining atoms to form molecules.

Functional groups Chemical patterns of atoms that exhibit consistent "function" (properties and reactivity) regardless of the specific molecule in which they appear.

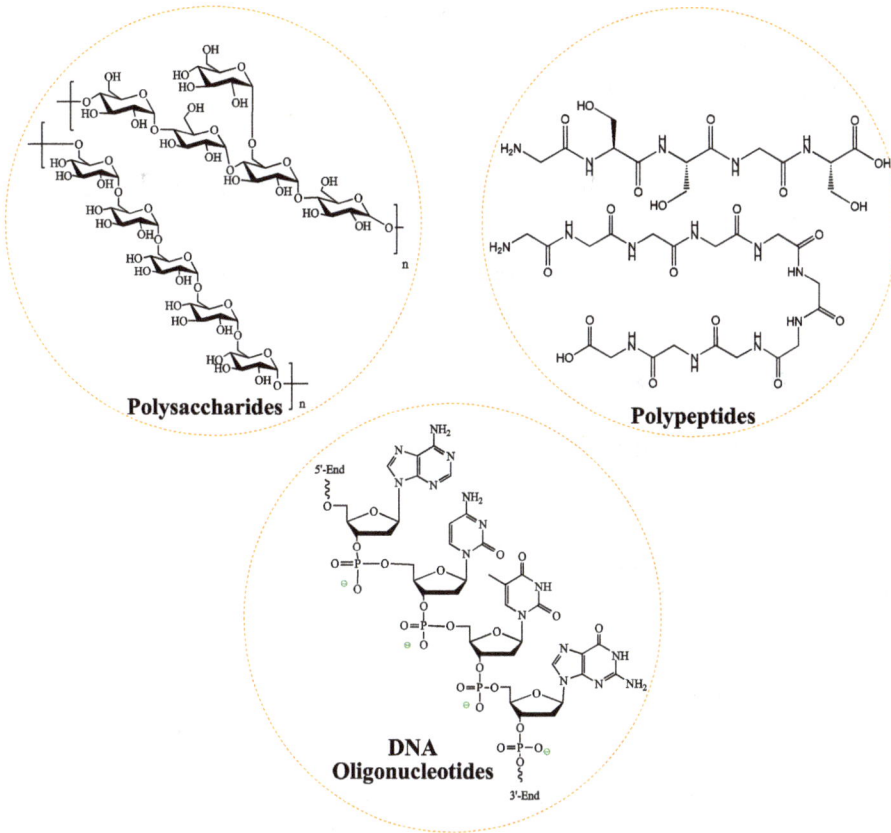

Figure 1.13: Selected examples of simplified biomacromolecule structures.

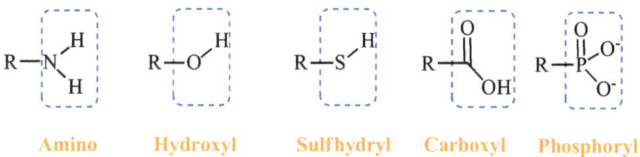

Amino   Hydroxyl   Sulfhydryl   Carboxyl   Phosphoryl

Figure 1.14: Selected functional groups in biomacromolecules.

O-**Glycosidic bond** Produced when the sugar's anomeric carbon bonds with the oxygen atom in the alcohol's hydroxyl group.

N-**Glycosidic bond** Occurs when the sugar's anomeric carbon links with the nitrogen atom of an amine.

Lipids A diverse group of organic compounds made from triglycerides which in turn formed from the esterification of glycerol with three fatty acids. Some lipids do not have fatty acids in their structure.

Nucleic acids Polymers of nucleotides. Their types are RNA and DNA.

Peptide bond An essential link that connects amino acids to form polypeptide chains, which fold into functional proteins.

Proteins Polymers of amino acids exist as small peptides or large polypeptides.

1.8 Practice Exercises

1.8.1 Briefly describe the covalent bond and give one example.
1.8.2 Complete the missing information in the following table:

Biomacromolecules	Monomers
Carbohydrates	1?
2?	Amino acids
Nucleic acids	3?

1.8.3 Why skeletal formula is preferable in drawing the structures of biomacromolecules?
1.8.4 Show how hydrogen bonds are formed between two molecules of HF.
1.8.5 What type of molecule is released during the peptide bond formation?
1.8.6 Name the following functional groups that play an important role in the structure and function of biomacromolecules:

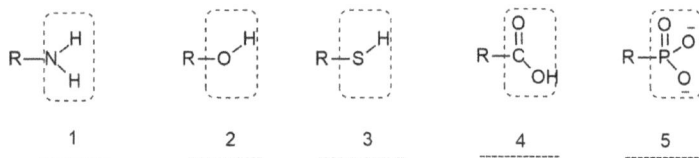

1.8.7 Show the hydrogen bond donor and hydrogen bond acceptor in the following hydrogen bonding scheme between two molecules of ammonia:

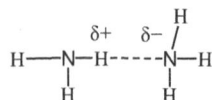

1.8.8 Show the O-glycosidic, N-glycosidic, and peptide bonds in the following structures:

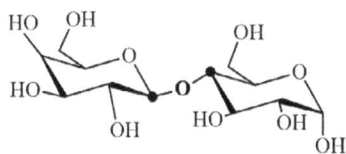

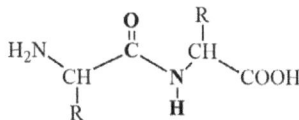

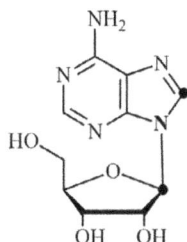

1.8.9 How many carbon atoms are there in the following sugars?
  i.   Pentose
  ii.   Tetrose
  iii. Triose

1.8.10 Define the hydrogen bond. Give two examples of molecules that can form hydrogen bonds.

1.8.11 What are the differences in electronegativities in the following molecules?
  i.   NaCl
  ii. Cl_2
  iii. H_2S

1.8.12 What are the commonly found functional groups in lipids?

1.8.13 What are the two types of natural nucleic acids?

1.8.14 Write down the full names of the following artificial nucleic acids:
  i.   MNA
  ii.   PNA
  iii. 4'-SDNA

1.8.15 What are the functional groups involved in the peptide bond formation?

1.8.16 Why are three fatty acids required for triglyceride formation?

1.8.17 What are the main atoms present in the structure of nucleic acids?

1.8.18 Why lipids are macromolecules but not polymers?

1.8.19 What are the monomeric units of proteins?

1.8.20 Give an example of dehydration reaction.

✳✳✳✳✳✳✳✳✳✳

Chapter 2
Carbohydrates

2.1 Definition

Carbohydrates are known as saccharides or, if small, sugars. They are polyhydroxy aldehydes or ketones and contain carbon, hydrogen, and oxygen atoms in their structures. The general formula is $C_x(H_2O)_y$ or $C_n(H_2O)_m$. To be classified as carbohydrates, molecules must have an aldehyde (CHO) or ketone (C=O) group as well as two or more hydroxyl (OH) groups. Figure 2.1 depicts the simplest carbohydrates: glyceraldehyde and dihydroxyacetone.

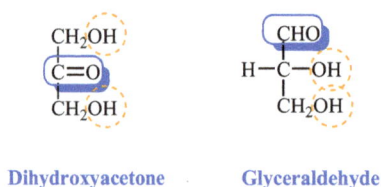

Dihydroxyacetone   **Glyceraldehyde**

Figure 2.1: Chemical structures of glyceraldehyde and dihydroxyacetone.

2.2 Classification

Several classifications of carbohydrates have proven to be useful with the most significant one stated as follows:

Carbohydrates are classified as simple or complex based on their complexity and the number of sugar units. Simple carbohydrates include monosaccharides, while complex carbohydrates include disaccharides, oligosaccharides, and polysaccharides.

–   Monosaccharide has only one single sugar unit:

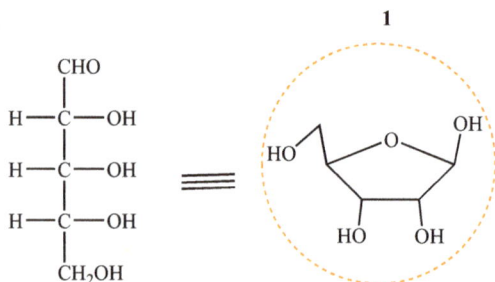

https://doi.org/10.1515/9783111583273-002

– Disaccharide has two sugar units:

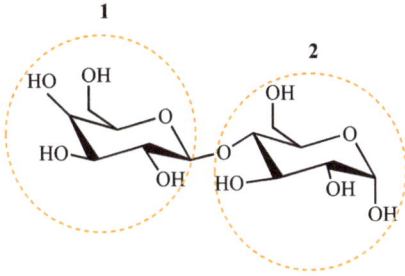

– Oligosaccharide has three or more sugar units:

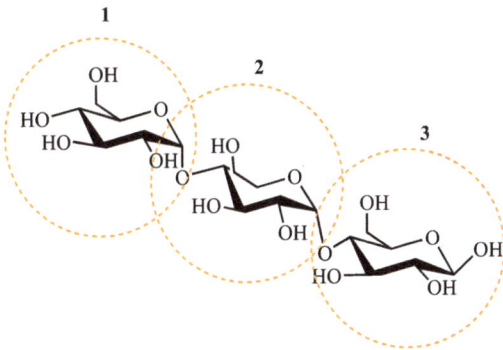

– Polysaccharide has at least 10 sugar units:

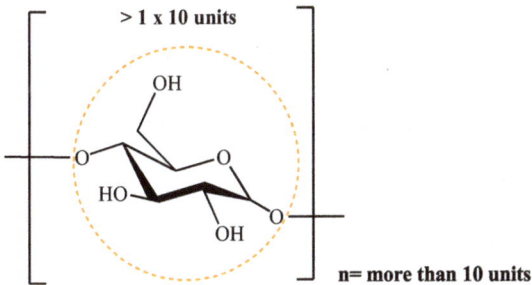

Carbohydrates can also be classified based on their reactivity into reducing sugars (*containing hemiacetal and in equilibrium with ring-opened form*) that are oxidized to carboxylic acids using standard test reagents like Benedict's reagent, Fehling's reagent, or Tollen's reagent. Nonreducing sugars (*containing acetals "locked" and are not in equilibrium with ring-opened form*) are those that are not oxidized by Tollen's or other reagents. Examples are shown in Figure 2.2.

Maltose is reducing sugar   Lactose is reducing sugar   Sucrose is non-reducing sugar
                                                                    "No hemiacetal"

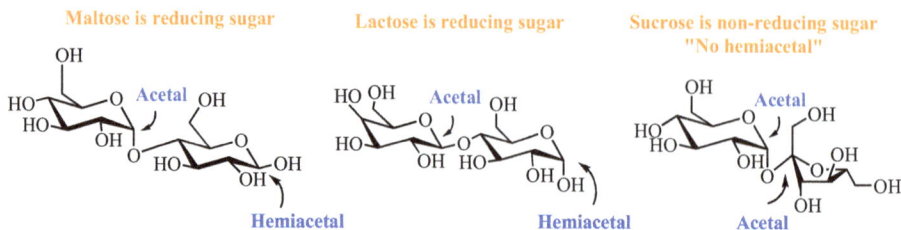

Hemiacetal   Hemiacetal   Acetal

Figure 2.2: Reducing and nonreducing sugars.

Monosaccharides can be further classified on the basis of number of carbon atoms in the structure or on the basis of carbonyl group location or on the basis of hydroxyl group orientation at the highest numbered chiral center.

Based on the counting of the number of carbon atoms present in the structure or on the size, monosaccharides can be classified into trioses (C3 sugars), tetroses (C4 sugars), pentoses (C5 sugars), and hexoses (C6 sugars), as shown in Figure 2.3.

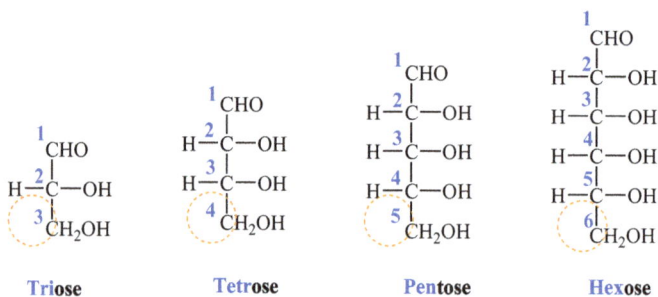

Triose   Tetrose   Pentose   Hexose

Figure 2.3: Chemical structures of triose, tetrose, pentose, and hexose.

Based on the type of carbonyl (C=O functional group) that exists in the structure, carbohydrates can be classified into aldoses or ketoses. An aldose has an aldehyde functional group and ketose has a ketone functional group, as shown in Figure 2.4.

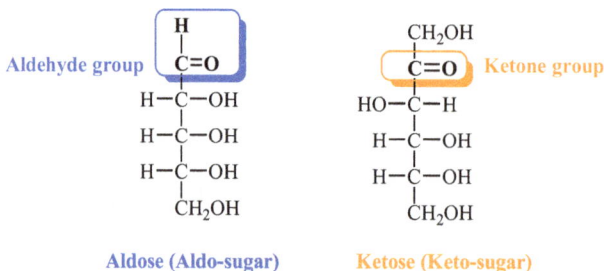

Aldose (Aldo-sugar)   Ketose (Keto-sugar)

Figure 2.4: Chemical structures of aldose and ketose.

The **D** and **L** designation of sugars is done based on their relationship to the glyceral-dehyde structure. The carbon chain is numbered starting at the carbonyl group end of the molecule, and the highest numbered chiral center is used to determine the **D** and **L** configurations, as presented in Figure 2.5.

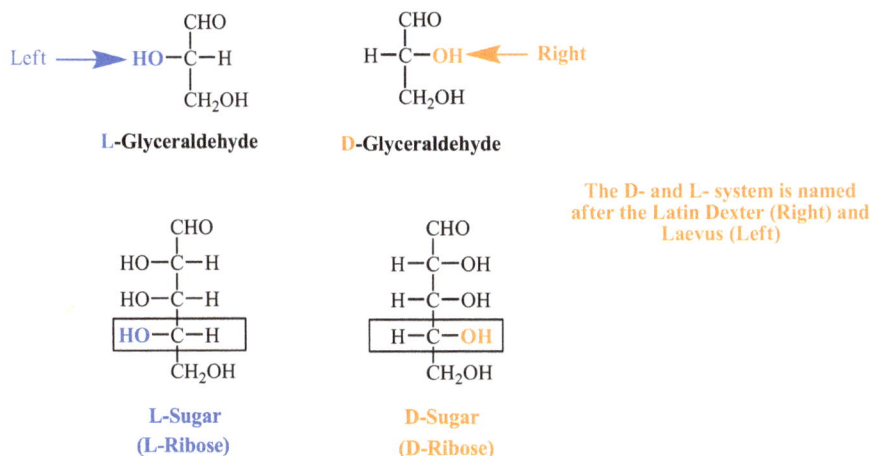

Figure 2.5: D and L configurations.

2.3 Monosaccharides

Monosaccharides include simple sugars and their derivatives. They are the fundamental carbohydrate units from which more complicated compounds are derived. Monosaccharides have carbon atoms with connected hydrogen atoms, at least two hydroxyl groups, and an aldehyde (RCHO) or ketone (RCOR) group. Monosaccharides' carbon number ranges from three to eight atoms, with the most frequent being five (e.g., pentoses, $C_5H_{10}O_5$) or six (e.g., hexoses, $C_6H_{12}O_6$). The monosaccharides are classified into two families: D-aldoses and D-ketoses (Figures 2.6 and 2.7).

2.3.1 Fischer, Wedge-Dash, Haworth, and Chair Forms of Monosaccharides

Monosaccharides are depicted or represented in various formats. These forms include the Fischer, wedge-dash, Haworth, and chair. The Fischer projection or form is used to depict stereochemistry in open-chain monosaccharides. Horizontal lines are thought to project out of the plane toward the reader, while vertical lines project behind the plane. A wedge and dash projection is a representation that depicts a molecule using three types of lines to describe its three-dimensional structure. Solid lines represent bonds on the paper's plane, whereas dashed lines represent bonds that reach away from the

CHO Aldehyde
H—C—OH Right
CH₂OH
D-Glyceraldehyde

CHO
H—C—OH
H—C—OH
CH₂OH
D-Erythrose

CHO
HO—C—H
H—C—OH
CH₂OH
D-Threose

CHO
HO—C—H
H—C—OH
H—C—OH
CH₂OH
D-Arabinose

CHO
H—C—OH
H—C—OH
H—C—OH
CH₂OH
D-Ribose

CHO
H—C—OH
HO—C—H
H—C—OH
CH₂OH
D-Xylose

CHO
HO—C—H
HO—C—H
H—C—OH
CH₂OH
D-Lycose

CHO
H—C—OH
H—C—OH
H—C—OH
H—C—OH
CH₂OH
D-Allose

CHO
HO—C—H
H—C—OH
H—C—OH
H—C—OH
CH₂OH
D-Altrose

CHO
HO—C—H
HO—C—H
H—C—OH
H—C—OH
CH₂OH
D-Mannose

CHO
H—C—OH
H—C—OH
HO—C—H
H—C—OH
CH₂OH
D-Gulose

CHO
HO—C—H
H—C—OH
HO—C—H
H—C—OH
CH₂OH
D-Idose

CHO
HO—C—H
HO—C—H
HO—C—H
H—C—OH
CH₂OH
D-Talose

CHO
H—C—OH
HO—C—H
H—C—OH
H—C—OH
CH₂OH
D-Glucose

CHO
H—C—OH
HO—C—H
HO—C—H
H—C—OH
CH₂OH
D-Galactose

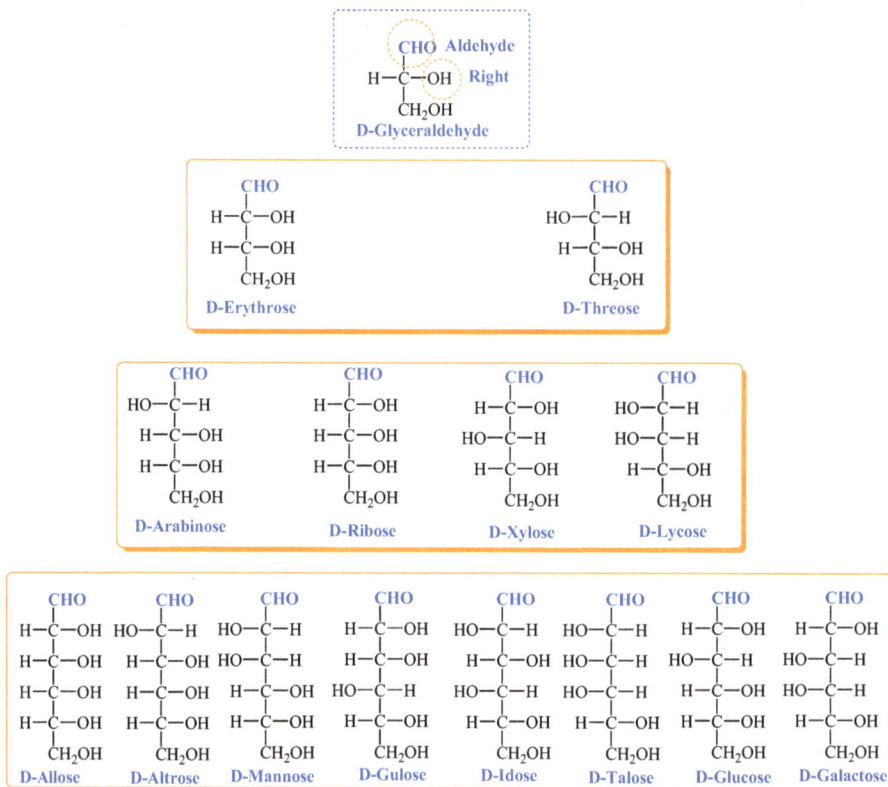

Figure 2.6: D-Aldose family.

viewer. Wedge-shaped lines represent bonds facing the viewer. The Haworth projection is a way of depicting cyclic monosaccharides (or sugars). The Haworth structure depicts the compound's bonds in its three-dimensional orientation. When a line points up, the bond is oriented up; when a line points down, the bond is oriented downward. Because many monosaccharides have the same constitutional structure as other monosaccharides and differ only in the 3D orientation of the bonds, understanding the bond orientation is critical when studying monosaccharides. Monosaccharides are also represented by the chair structure. Figure 2.8 depicts all of these forms.

2.3.2 Cyclization (Ring Formation) of Monosaccharides

2.3.2.1 Cyclization of D-Glucose

Cyclization occurs in two consecutive steps. In step 1, the linear aldehyde of D-glucose is turned on its side, and rotation about the C4–C5 bond puts the C5-hydroxyl function

CH$_2$OH
|
C=O   Ketone
|
CH$_2$OH
Dihydroxyacetone

CH$_2$OH
|
C=O
|
H—C—OH   Right
|
CH$_2$OH
D-Erythrulose

CH$_2$OH
|
C=O
|
H—C—OH
|
H—C—OH
|
CH$_2$OH
D-Ribulose

CH$_2$OH
|
C=O
|
HO—C—H
|
H—C—OH
|
CH$_2$OH
D-Xylulose

CH$_2$OH
|
C=O
|
H—C—OH
|
H—C—OH
|
H—C—OH
|
CH$_2$OH
D-Psicose

CH$_2$OH
|
C=O
|
HO—C—H
|
H—C—OH
|
H—C—OH
|
CH$_2$OH
D-Fructose

CH$_2$OH
|
C=O
|
H—C—OH
|
HO—C—H
|
H—C—OH
|
CH$_2$OH
D-Sorbose

CH$_2$OH
|
C=O
|
HO—C—H
|
HO—C—H
|
H—C—OH
|
CH$_2$OH
D-Tagatose

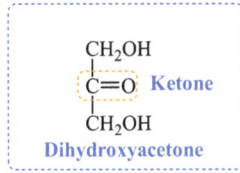

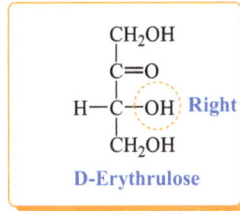

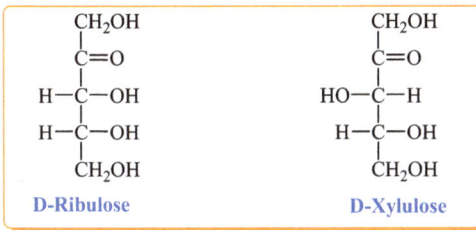

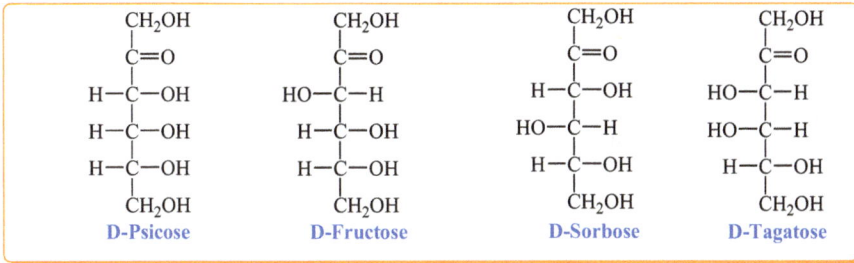

Figure 2.7: D-Ketose family.

closer to the aldehyde carbon. In step 2, the attack of the OH group linked to C-5 on C=O results in the production of α- and β-D-glucopyranose (Figure 2.9).

2.3.2.2 Cyclization of D-Fructose

Cyclization proceeds in two steps. In step 1, the linear aldehyde of the **D**-fructose is tipped on its side, and rotation about the C4–C5 bond brings the C5-hydroxyl function close to the aldehyde carbon. While in step 2, the attack by OH group attached to C-5 on C=O will lead to the formation of α- and β-**D**-fructofuranose (Figure 2.10).

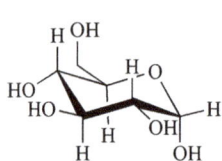

Chair Conformation

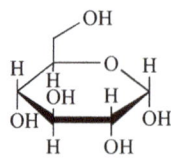

Haworth Projection

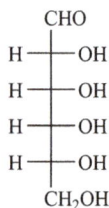

Fischer Projection

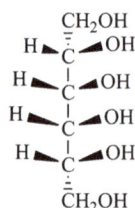

Wedge and Dash Projection

Figure 2.8: Structural forms of monosaccharides.

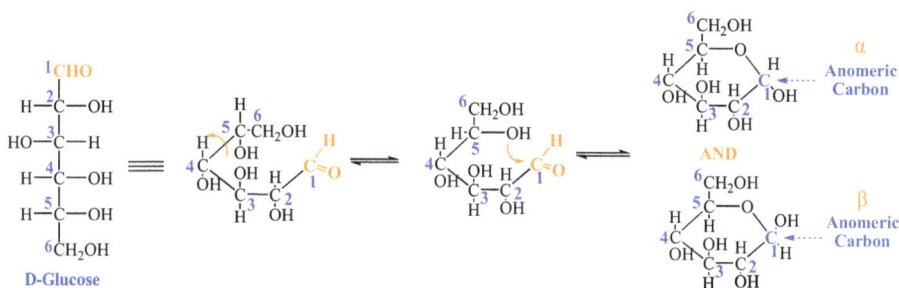

Figure 2.9: Cyclization of D-glucose.

2.3.2.3 Cyclization of D-Ribose

Cyclization proceeds in two steps. In step 1, the linear aldehyde of **D**-ribose is tipped on its side, and rotation about the C3–C4 bond brings the C4-hydroxyl function close to the aldehyde carbon. While in step 2, the attack by OH group attached to C-4 on C=O will lead to the formation of α- and β-**D**-ribofuranose (Figure 2.11).

2.3.3 Reactions of Monosaccharides

2.3.3.1 Glycosidation (Glycoside Bond Formation)

A glycosidic bond is a functional group that joins a carbohydrate (sugar) molecule to an alcohol, which may be another carbohydrate. A glycosidic bond is created between

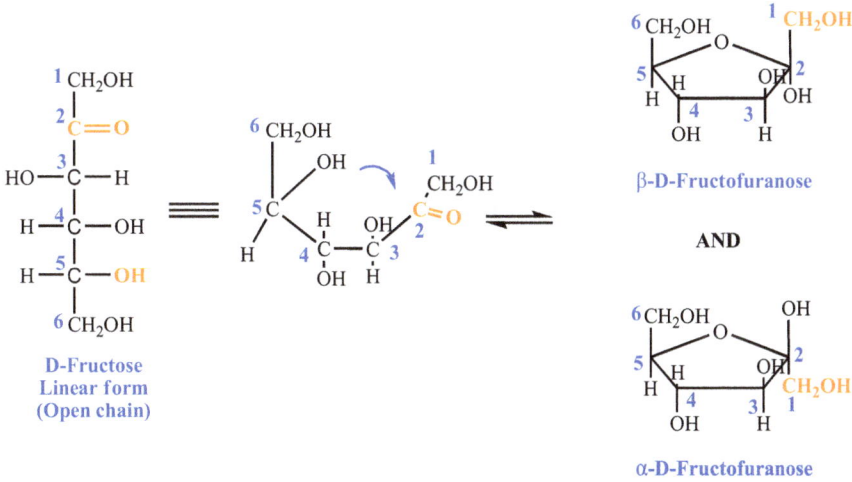

Figure 2.10: Cyclization of D-fructose.

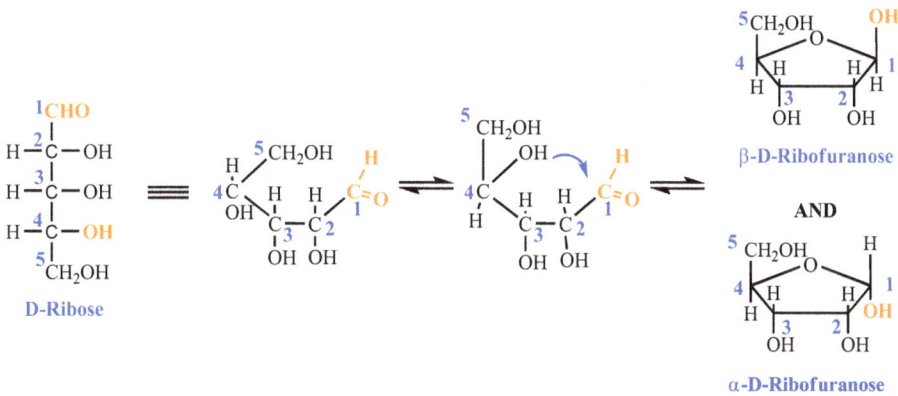

Figure 2.11: Cyclization of D-ribose.

the hemiacetal group of a saccharide (or a saccharide-derived molecule) and an alcohol's hydroxyl group. Glycosides are acetals found on the anomeric carbon of carbohydrates. When glucose reacts with an alcohol (such as methanol) in the presence of catalytic acid, methyl glucoside is produced, as shown in Figure 2.12.

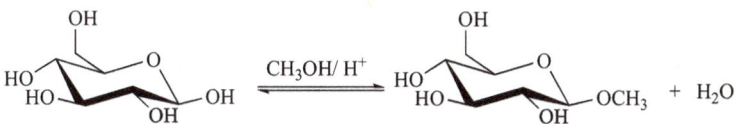

Figure 2.12: Methyl glucoside formation.

2.3.3.2 Hydrolysis of Glycosides

When glycosides are subjected to acidic conditions, hydrolysis takes place, leading to a mixture of anomers as shown in Figure 2.13.

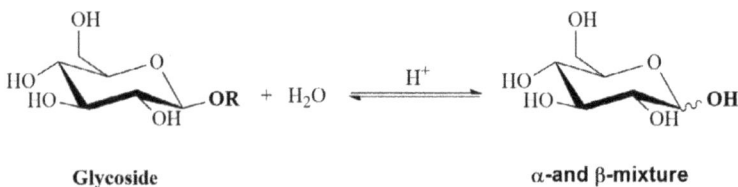

Figure 2.13: Hydrolysis of glycosides.

Disaccharides can be readily hydrolyzed under weak acidic conditions, such as dilute HCl, producing their constitutive monomers in equivalent quantities (Figure 2.14).

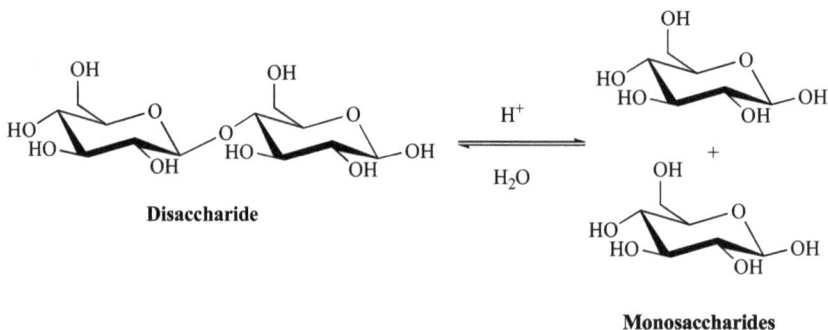

Figure 2.14: Hydrolysis of disaccharides.

2.3.3.3 Alkylation of Glycosides (Formation of Ethers)

When glucoside reacts with methyl iodide in the presence of silver oxide, the methylated glycoside is obtained (Figure 2.15).

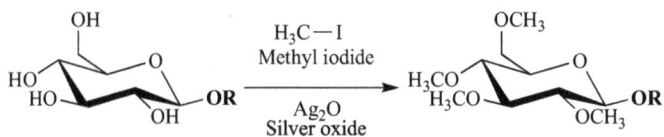

Figure 2.15: Formation of ethers (alkylation of glycosides).

2.3.3.4 Acylation of Glycosides (Formation of Esters)

When glucoside reacts with acetic anhydride in the presence of pyridine, the acetylated glycoside is obtained (Figure 2.16).

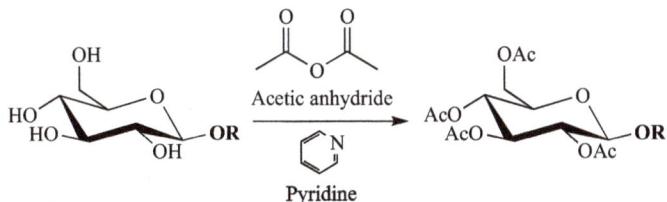

Figure 2.16: Formation of esters (acylation of glycosides).

2.3.3.5 Oxidation to Acidic Sugars

Oxidation of the aldehyde end of D-glucose with a weak oxidizing agent such as bromine water produces **D-gluconic acid** (Figure 2.17).

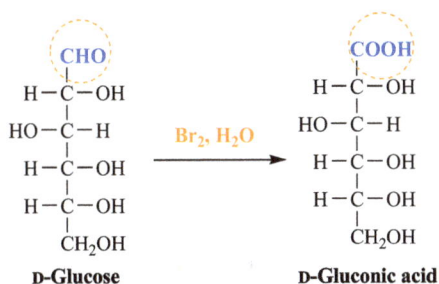

Figure 2.17: Oxidation of D-glucose to D-gluconic acid.

2.3.3.6 Reduction to Sugar Alcohols

The reduction of the carbonyl group in either an aldose or a ketose to a hydroxyl group using hydrogen as the reducing agent or sodium borohydride produces the corresponding polyhydroxy alcohol, which is sometimes called a *sugar alcohol*. For example, the reduction of **D-glucose** gives **D-glucitol** (Figure 2.18).

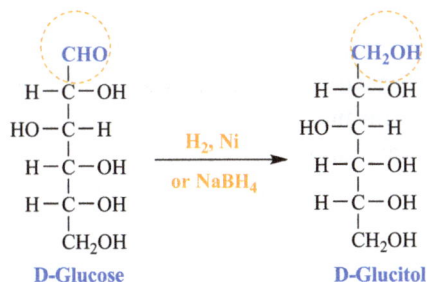

Figure 2.18: Reduction of D-glucose to D-glucitol.

2.3.3.7 Koenigs-Knorr Reaction

The presence of multiple –OH groups on a sugar molecule makes glycoside synthesis extremely challenging. The Koenigs-Knorr reaction, one of the oldest glycosidation reactions, is effective for generating glucose β-glycosides. The process begins by reacting glucose pentaacetate with HBr, which produces pyranosyl bromide. The presence of silver oxide allows for the nucleophilic addition of the chosen alcohol. The hydrolysis of the remaining acetal groups under basic conditions yields phenyl-β-glucopyranoside (Figure 2.19).

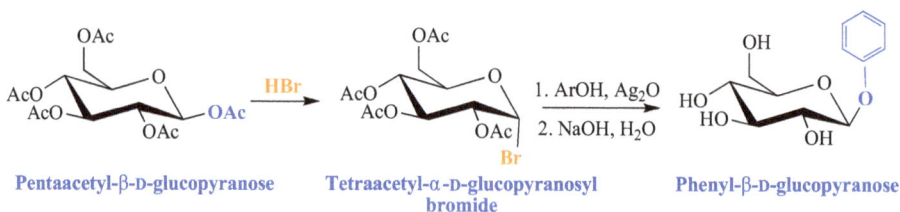

Pentaacetyl-β-D-glucopyranose   Tetraacetyl-α-D-glucopyranosyl bromide   Phenyl-β-D-glucopyranose

Figure 2.19: Koenigs-Knorr reaction.

2.3.3.8 Phosphorylation Reaction

The most common ionic group in biologically significant organic molecules is phosphate; hence, phosphorylation of alcohol groups is a crucial metabolic process. In glucose phosphorylation, ATP is the phosphate donor, and the mechanism is highly consistent: alcohol oxygen in glucose works as a nucleophile, attacking ATP's gamma-phosphorus and releasing ADP (Figure 2.20).

2.3.3.9 Chain Lengthening: The Kiliani-Fischer Synthesis

The Kiliani-Fischer synthesis extends the carbon chain of an aldose by one carbon atom. In doing so, two new sugar epimers are formed. The synthesis begins by reacting an aldose with HCN. Nucleophilic addition adds the cyanide nucleophile to the aldose aldehyde's electrophilic carbon, resulting in the formation of a cyanohydrin intermediate. The cyanide nucleophile incorporates carbon. Stereochemical control is lost, resulting in a racemic mixture of two cyanohydrins with distinct stereochemistry at C2. The cyanohydrin's nitrile group is transformed into an imine intermediate by hydrogenation over a palladium catalyst. Finally, the imine is hydrolyzed into an aldehyde, resulting in two new aldoses with additional carbon. For example, performing the Kiliani-Fischer synthesis on D-ribose produces a mixture of D-allose and D-altrose (Figure 2.21).

2.3.3.10 Chain Shortening: The Wohl Degradation

Wohl degradation cleaves an aldose chain's C1–C2 link, shortening it by one carbon. The aldose aldehyde is transformed into an oxime by treating it with hydroxylamine

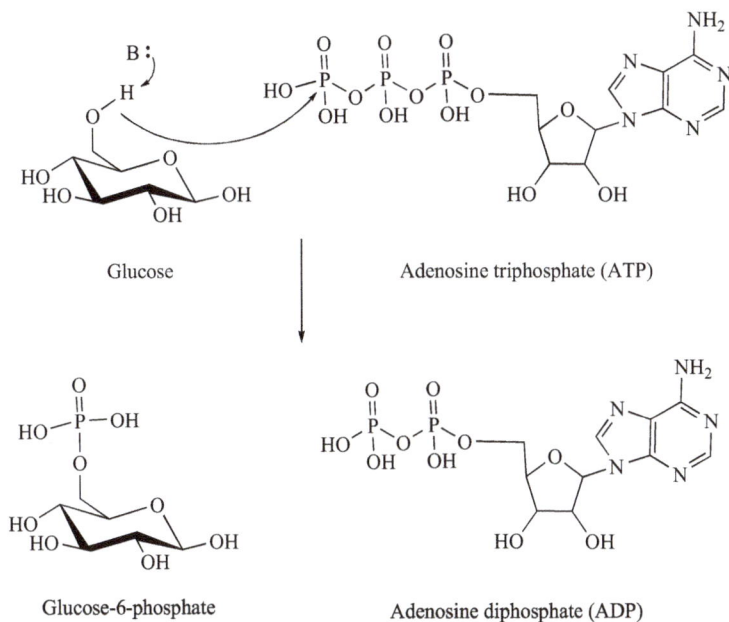

Figure 2.20: Phosphorylation of glucose.

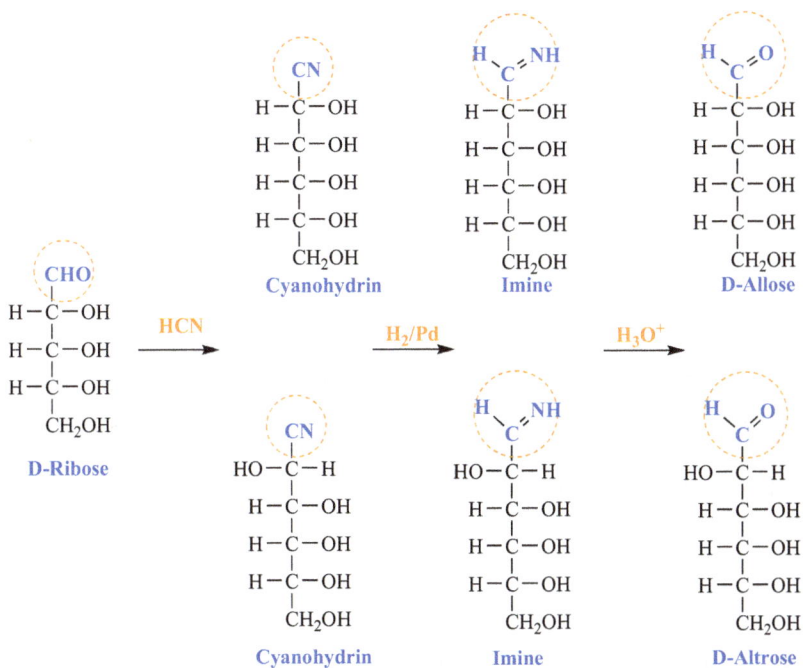

Figure 2.21: The Kiliani-Fischer synthesis.

(NH$_2$OH). Dehydration of oxime with acetic anhydride yields cyanohydrin. Under basic conditions, cyanohydrin loses HCN and forms an aldehyde carbonyl. The examples below demonstrate how to use this process to shorten D-glucose to D-arabinose and D-arabinose to generate D-erythrose (Figure 2.22).

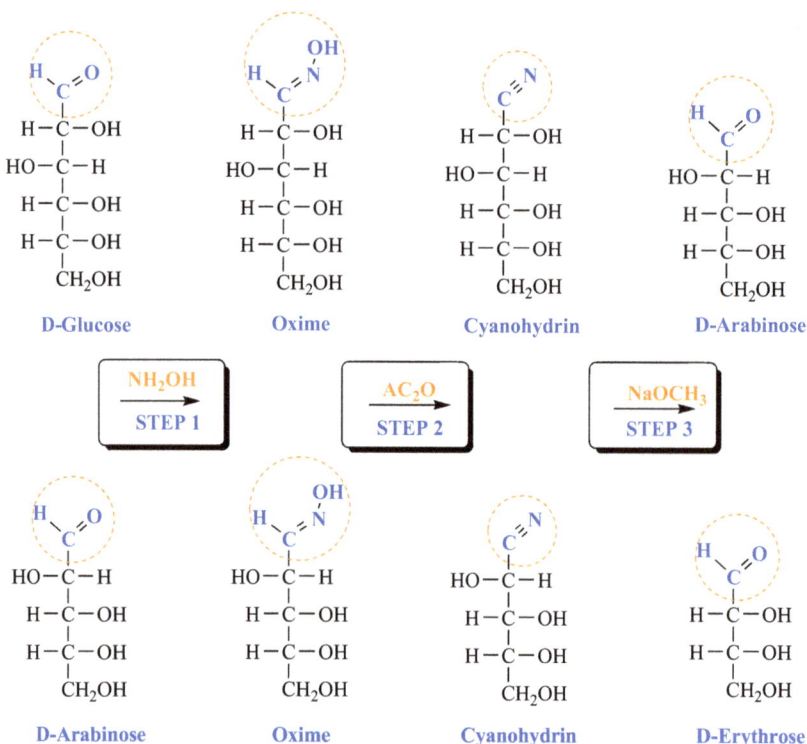

Figure 2.22: The Wohl degradation.

2.3.4 Mutarotation

Mutarotation is the change in equilibrium between two anomers due to optical rotation. Cyclic sugars show mutarotation as α and β anomeric forms interconvert. For example, an aqueous solution of D-glucose contains an equilibrium mixture of α-D-glucopyranose, β-D-glucopyranose, and the intermediate open-chain form (Figure 2.23).

2.3.5 Epimerization

Epimerization is a process in carbohydrate stereochemistry in which there is a change in the configuration of only one chiral center. For example, **C2**-epimerization occurs

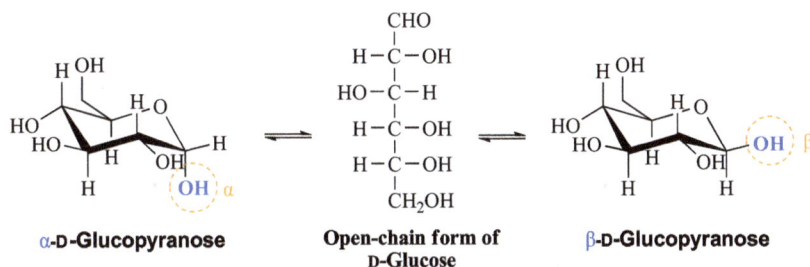

Figure 2.23: Mutarotation of D-glucose.

between D-glucose and D-mannose and **C4**-epimerization occurs between D-glucose and D-galactose (Figure 2.24).

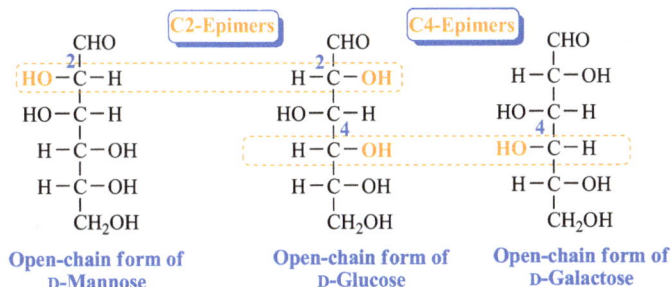

Figure 2.24: C2-Epimerization between D-glucose and D-mannose and C4-epimerization between D-glucose and D-galactose.

2.3.6 Isomerization (Enolization-Tautomerization) of D-Glucose to D-Fructose

D-Glucose undergoes *isomerization* to form D-fructose by the action of an alkaline solution. This is an overall reaction that involves two steps, namely *enolization* and *tautomerization*.

An alkaline solution causes D-glucose to isomerize into D-fructose. This overall reaction consists of two steps: enolization and tautomerization. The base eliminates the proton (alpha hydrogen) next to the anomeric carbonyl carbon in this isomerization. As one of the C=O (carbonyl) bonds breaks, a double bond forms between the alpha carbon and the carbonyl carbon, and the freed electron pair on oxygen accepts a proton from an acid. This results in the transfer of a proton from the alpha location to carbonyl oxygen. Further rearrangement of the enediol results in the creation of fructose ketones. In summary, the glucose aldehyde is enolized, as shown in Figure 2.25, and then tautomerized to provide the fructose ketone.

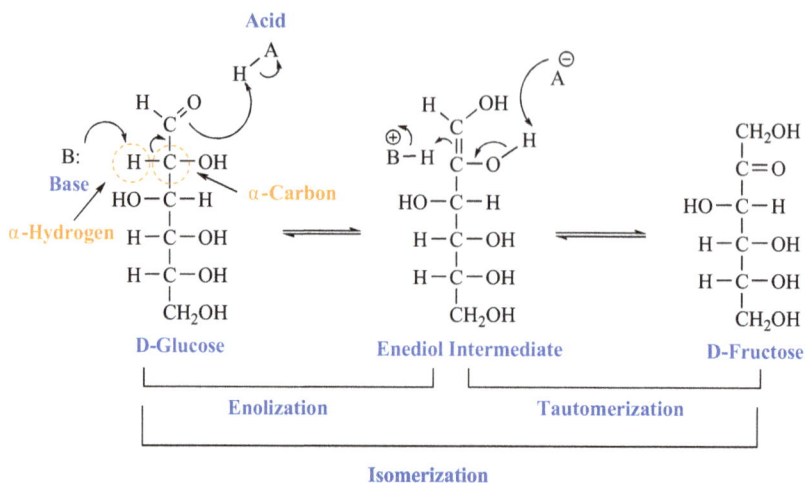

Figure 2.25: Isomerization of D-glucose to D-fructose.

2.4 Disaccharides

A disaccharide, also known as a double sugar or biose, is the sugar made when two monosaccharides are joined together by a glycosidic linkage. In the following chemical structures, the sugar rings are designated **A** and **B**, and the glycoside bond is circled in **blue**. Maltose, isomaltose, sucrose, and trehalose have alpha glycoside bonds, whereas lactose, cellobiose, and gentiobiose include beta glycoside bonds.

2.4.1 Lactose

Lactose is a disaccharide consisting of galactose and glucose units connected by a $\beta(1\rightarrow4)$ glycosidic bond (Figure 2.26). Lactose is present in milk and dairy products. Lactose can be classified as an added or natural sugar, depending on where it comes from. Lactose is naturally present in mammalian milk and can be isolated and crystallized for use in baked goods, caramels, frozen sweets, fudge, meat products, sauces, and soups. Lactose, added as a manufacturing element to packaged goods and beverages, is referred to as added sugar, but lactose found in basic milk and plain milk products is considered a natural sugar.

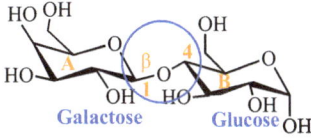

Figure 2.26: Chemical structure of lactose.

2.4.2 Sucrose

Sucrose, or saccharose, is a disaccharide made up of fructose and glucose units joined by α(1→2) glycosidic bond (see Figure 2.27). It is commonly referred to as "table sugar," yet it is found naturally in fruits, vegetables, and nuts. However, it is also economically produced by refining sugarcane and sugar beet.

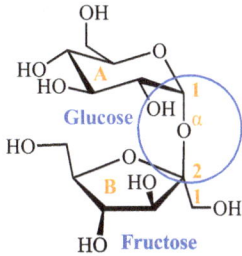

Figure 2.27: Chemical structure of sucrose.

2.4.3 Maltose

Maltose, commonly known as maltobiose or malt sugar, is a disaccharide made from two units of glucose connected by an α(1→4) glycosidic bond between the glucose units. It is a two-unit member of the amylose homologous series, the principal structural component of starch (Figure 2.28).

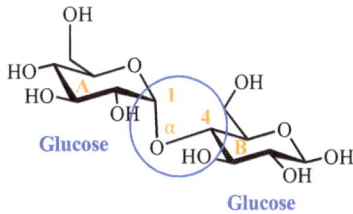

Figure 2.28: Chemical structure of maltose.

2.4.4 Isomaltose

This is a glycosylglucose consisting of two D-glucose units (Figure 2.29). It is similar to maltose but with α(1→6) glycosidic bond instead of the α(1→4) glycosidic bond. Due to that isomaltose can have different digestive properties when compared to other sugars like maltose, sucrose, or glucose. This particular relationship can alter how the body processes and absorbs isomaltose, influencing the glycemic response. Isomaltose is often used as a substitute for sucrose in a variety of sugar-free and low-sugar products, allowing diabetics to consume sweet-tasting meals and beverages without dramatically raising their blood sugar levels.

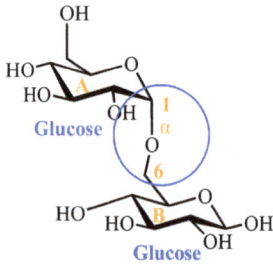

Figure 2.29: Chemical structure of isomaltose.

2.4.5 Cellobiose

A disaccharide with two units of glucose is joined by a β(1→4) glycosidic bond (Figure 2.30) and can be produced by the hydrolysis of cellulose.

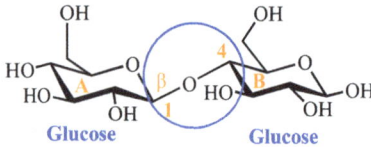

Figure 2.30: Chemical structure of cellobiose.

2.4.6 Gentiobiose

Gentiobiose is a disaccharide composed of two D-glucose units connected by a β(1→6) glycosidic bond (Figure 2.31). It is a white, crystalline material that dissolves in water or hot methanol. Gentiobiose is incorporated into the molecular structure of Crocin, the molecule that gives saffron its color.

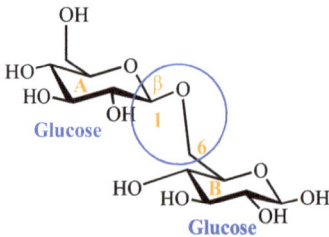

Figure 2.31: Chemical structure of gentiobiose.

2.4.7 Trehalose

Trehalose is a nonreducing, natural disaccharide composed of two glucose units connected together by α(1→1) glycosidic bond (Figure 2.32). It is found in various organisms, including fungi, bacteria, and invertebrates. Trehalose is a bis-acetal, and is
therefore a nonreducing sugar. Trehalose is well-known as an important protein stabilizer and reduces corneal damage of the eye caused by ultraviolet B radiation.

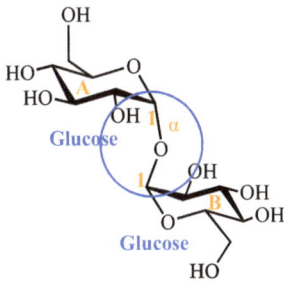

Figure 2.32: Chemical structure of trehalose.

2.5 Oligosaccharides

Oligosaccharides are saccharide molecules composed of 2–10 monosaccharide units
formed through glycosidic bond polymerization. Many oligosaccharides are found
naturally in common fruits and vegetables. The human digestive tract is unable to
break down the vast majority of oligosaccharides. Instead, they move down the digestive tract and into the colon, where they nourish and support the growth of beneficial
bacteria. As a result of this, oligosaccharides are classed as prebiotics.

2.5.1 Raffinose

Raffinose or melitose is a trisaccharide that consists of a galactose unit coupled to sucrose [glucose connected to fructose by α(1→2)] by α(1→6) glycosidic bond (Figure 2.33).
The enzyme α-galactosidase (α-GAL) can hydrolyze the nondigestible short-chain raffinose to produce D-galactose and sucrose.

2.5.2 Maltotriose

Maltotriose is a trisaccharide made up of three glucose units connected by α (1, 4) linkages (Figure 2.34). It is the primary fermentable sugar in the wort. It is formed during
mashing as a result of the enzymatic breakdown of starches.

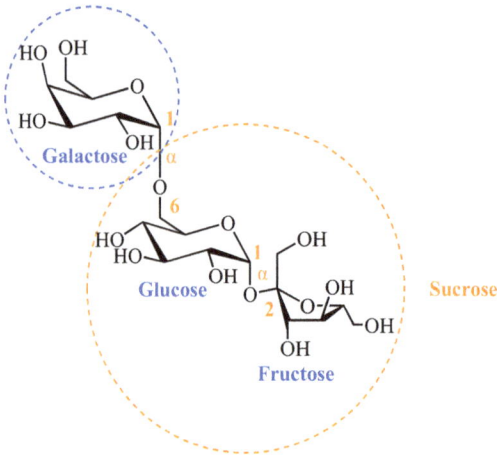

Figure 2.33: Chemical structure of raffinose.

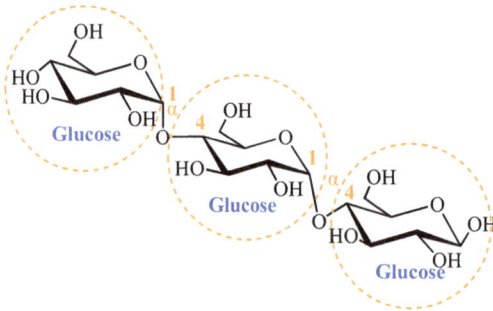

Figure 2.34: Chemical structure of maltotriose.

2.5.3 Stachyose

Stachyose is a tetrasaccharide that consists of two α-D-galactose units, one α-D-glucose unit, and one β-D-fructose unit sequentially linked as galactose, galactose, glucose, and fructose [α(1→6), α(1→6), α(1→2)] (Figure 2.35).

2.5.4 Maltodextrins

Maltodextrins are composed of D-glucose units linked in chains of varying lengths (3–20 sugar units). The glucose units are typically connected by α(1→4) glycosidic bonds. Because of its chain length, maltodextrin is classified as an oligosaccharide (if $n = 1$, Figure 2.36) and a polysaccharide (if $n = 18$, Figure 2.36). It is commonly used in foods

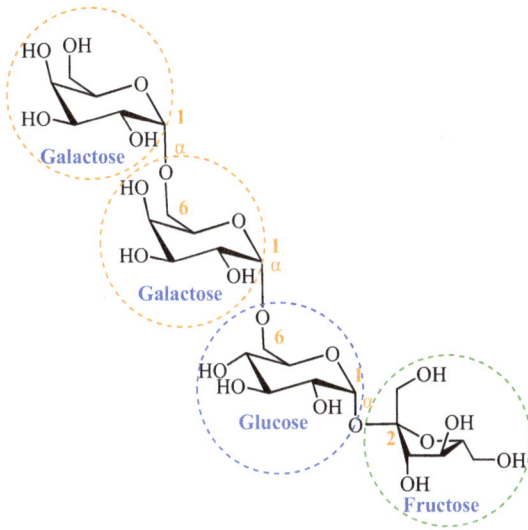

Figure 2.35: Chemical structure of stachyose.

and beverages as a thickening, sweetener, and/or stabilizer. It is a relatively short-chain polymer formed by partially hydrolyzing grain starches, usually corn or wheat. Maltodextrin is a popular food component in a variety of products, including infant formula, ice cream, salad dressing, and peanut butter, due to its safety, low cost, and high-water solubility.

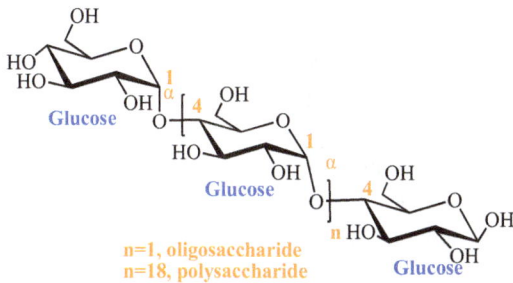

Figure 2.36: Chemical structure of maltodextrins.

2.5.5 Maltotetraose

Maltotetraose, a tetrasaccharide composed of glucose molecules connected by α-1,4 glycosidic bonds, is present in *Bacillus stearothermophilus* (Figure 2.37).

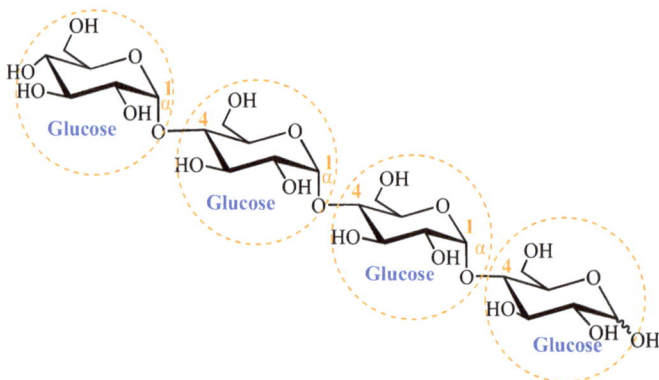

Figure 2.37: Chemical structure of maltotetraose.

2.6 Polysaccharides

Polysaccharides are polymers made up of multiple monosaccharide units (**>10 sugar units**) held together by glycosidic linkages. The type of monosaccharide repeating units in the polymer chain, its length, the type of glycosidic linkage between monomer units, and the degree of polymer chain branching make it simple to distinguish between them.

2.6.1 Starch

Starch is made of two polymers such as amylose and amylopectin, both of which are made from α-D-glucose connected together by α(1→4) glycosidic bonds (Figure 2.38). *Amylose* is a linear polysaccharide made up of about 500–20,000 glucose units. It is soluble in water and can be digested into glucose units by the enzyme α-amylase. Amylase is also responsible for the intense blue color observed in the presence of iodine when it becomes caught in the amylose helices. *Amylopectin* is a highly branched polysaccharide made up of about 2,000–200,000 glucose units. Its short side chains are connected by α(1→6) glycosidic bonds and the branching points appear every 20–30 sugar units. Amylopectin is a water-insoluble polymer and accounts for 80% of all starch.

Crushed or ground starch-containing tubers or seeds are used to make commercial starch, and the pulp is then combined with water. To get rid of any leftover contaminants, the resultant paste is dried. Paper bags, boxes, gummed paper, tape, and corrugated paperboard are all made with starch. The textile industry also uses a lot of starch for warp sizing, which gives strength to the thread during weaving.

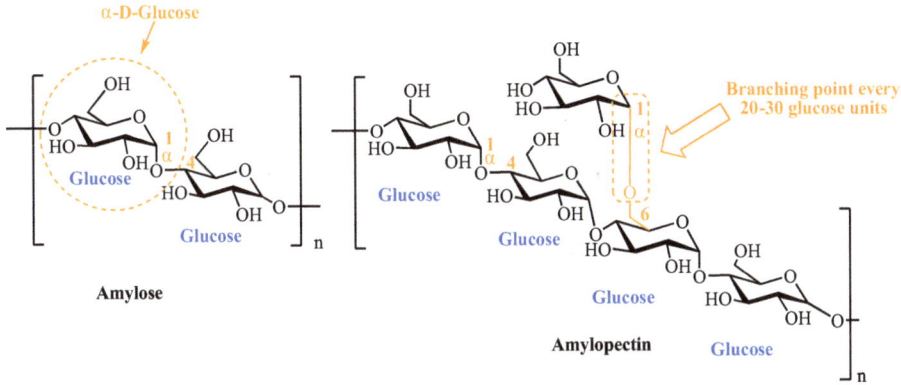

Figure 2.38: Simplified chemical structures of amylose and amylopectin.

2.6.2 Cellulose

Cellulose is a polymer made up of β-D-glucose units connected together by β(1→4) gly-cosidic bonds. The hydroxymethyl (CH$_2$OH) groups alternate above and below the plane of the cellulose molecule, forming long unbranched chains (Figure 2.39). It is available in three different forms: microcrystalline cellulose, powdered cellulose (PC), and low-crystallinity PC.

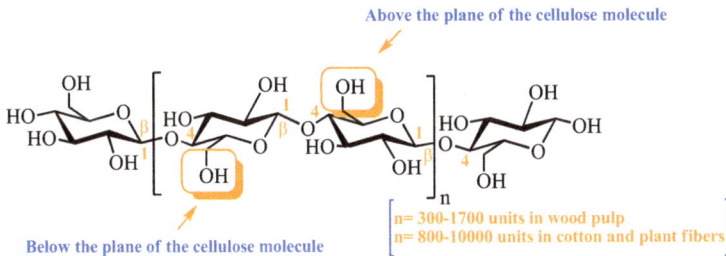

Figure 2.39: Simplified chemical structure of cellulose.

There are various cellulose derivatives (Figure 2.40). These include cellulose esters and cellulose ethers. The ester polymers are water-insoluble and commonly employed in pharmaceutical controlled-release formulations. They are divided into organic and inorganic groups. Examples of organic cellulose esters include cellulose acetate and cellulose acetate phthalate. Examples of inorganic cellulose esters are cellulose nitrate (also known as pyroxylin) and cellulose sulfate.

Cellulose ethers (Figure 2.41) are high-molecular-weight compounds produced by replacing the hydrogen atoms of hydroxyl groups with alkyl or substituted alkyl groups. Examples of mostly used cellulose ethers are methylcellulose, ethyl cellulose, hydroxye-

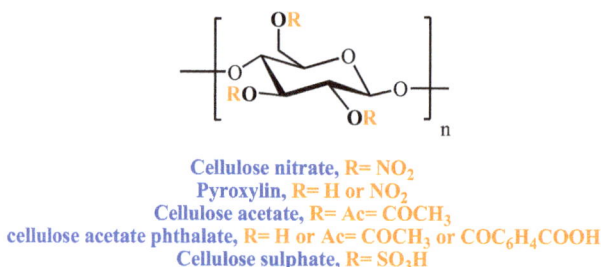

Cellulose nitrate, R= NO_2
Pyroxylin, R= H or NO_2
Cellulose acetate, R= Ac= $COCH_3$
cellulose acetate phthalate, R= H or Ac= $COCH_3$ or COC_6H_4COOH
Cellulose sulphate, R= SO_3H

Figure 2.40: Cellulose esters.

thylcellulose, hydroxypropyl cellulose, hydroxypropylmethylcellulose, carboxymethyl-cellulose, and sodium carboxymethyl cellulose.

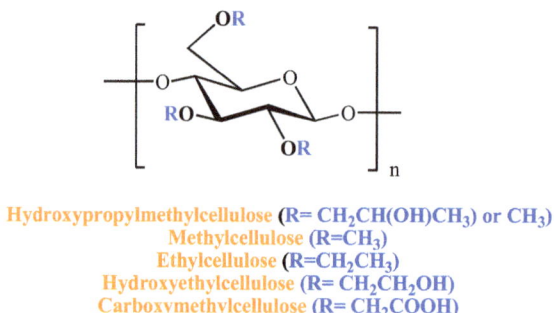

Hydroxypropylmethylcellulose (R= $CH_2CH(OH)CH_3$) or CH_3)
Methylcellulose (R=CH_3)
Ethylcellulose (R=CH_2CH_3)
Hydroxyethylcellulose (R= CH_2CH_2OH)
Carboxymethylcellulose (R= CH_2COOH)

Figure 2.41: Cellulose ethers.

2.6.3 Pullulan

Pullulan consists of maltotriose (three glucose molecules linked by α(1→4) glycosidic bonds) units linked through α(1→6) glycosidic bonds (Figure 2.42). Pullulan is water-soluble, odorless, flavorless, and edible, and makes strong films with high adhesion and oxygen barrier properties. Its films have low permeability to oxygen, which protects active ingredients, flavors, and colors incorporated into the film from deterioration. Pullulan is used as a food ingredient and coating agent in the food formulation and packaging industry.

2.6.4 Glycogen

Glycogen is a polymer of α-D-glucose. It may consist of an average of 2,000–60,000 glucose units connected together by α(1→4) glycosidic bonds. The branches in glycogen tend to be

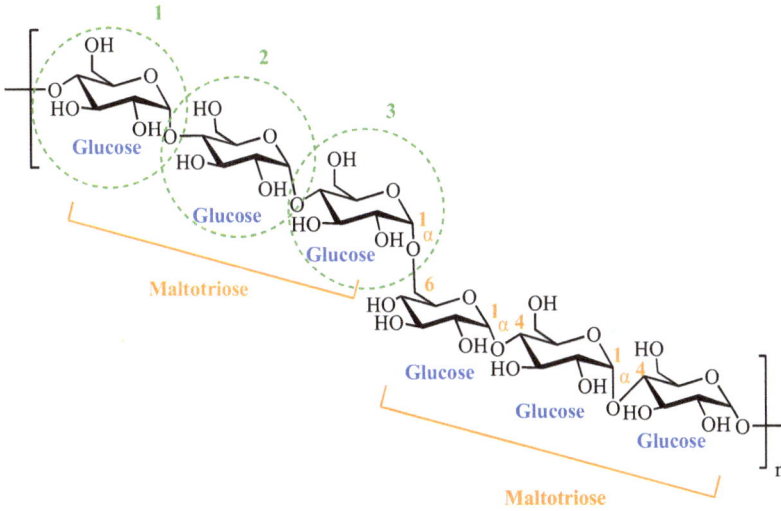

Figure 2.42: Simplified chemical structure of pullulan.

about 10–15 glucose units connected through α(1→6) glycosidic bonds (Figure 2.43). Glycogen acts as an energy reserve for the organism. A sudden lack of fuel would cause significant problems for cell activities and brain cells, and the body retains a reserve supply. As the blood glucose levels drop, the liver turns glycogen into glucose and delivers it into the bloodstream.

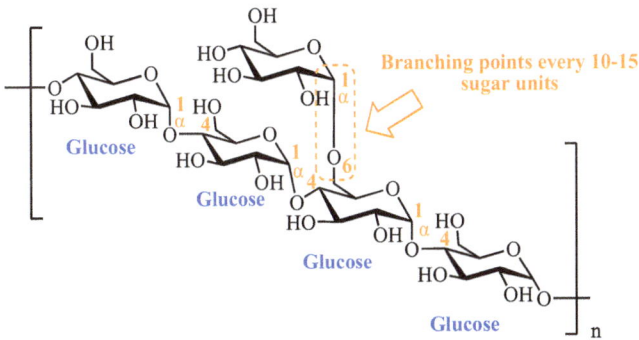

Figure 2.43: Simplified chemical structure of glycogen.

2.6.5 Dextran

Dextran is a high-molecular-weight polysaccharide formed by α(1→6) linked glucose units, with α(1→3) branch linkages and may contain other branch linkages such as α(1→2) or α(1→4). At least 50–60% of the linkage must be α(1→6) to define the molecule as dextran (Figure 2.44). The chains can consist of approximately 200,000 glucose units.

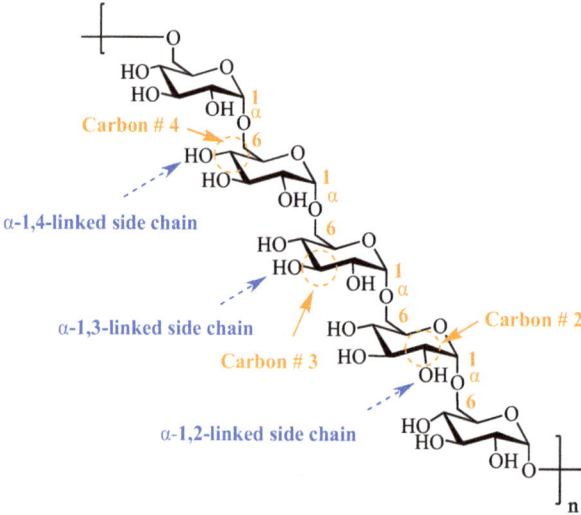

Figure 2.44: Simplified chemical structure of dextran.

2.6.6 Inulin

Inulin is a water-soluble polysaccharide and its polymer consists of fructose units linked via β(2→1) linkages that have a terminal glucose unit (Figure 2.45).

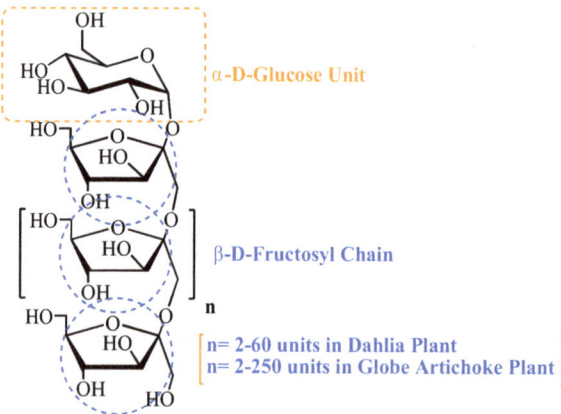

Figure 2.45: Simplified chemical structure of inulin.

2.6.7 Chitin

Chitin is a polysaccharide that is similar to cellulose in both function and structure. Structurally, chitin is a linear polymer (no branching) with all $\beta(1\rightarrow4)$ glycosidic linkages, as is cellulose. Chitin differs from cellulose in that the monosaccharide present is an *N*-acetyl amino derivative of D-glucose (Figure 2.46).

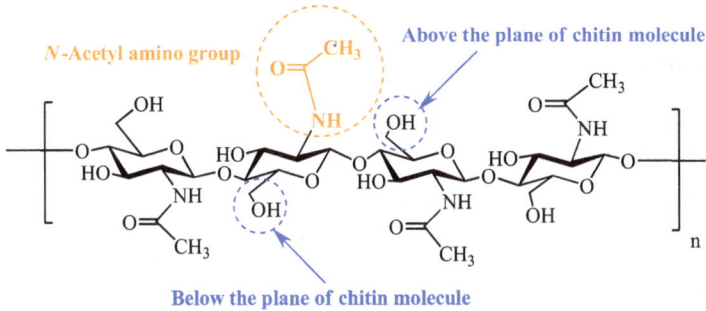

Figure 2.46: Simplified chemical structure of chitin.

2.6.8 Lignin

Lignin is a complex, oxygen-containing macromolecule with an irregular phenolic structure (Figure 2.47). It does not consist of any carbohydrate monomers. Lignin and cellulose work together to offer structural support in plants. It protects the cellulose structure against microbial attack. Lignin is a long-lasting, waterproof macromolecule that serves as the "backbone" of plants, providing structure and stability. It works as a barrier, preventing water from leaving a plant in dry conditions and repelling insects and fungi.

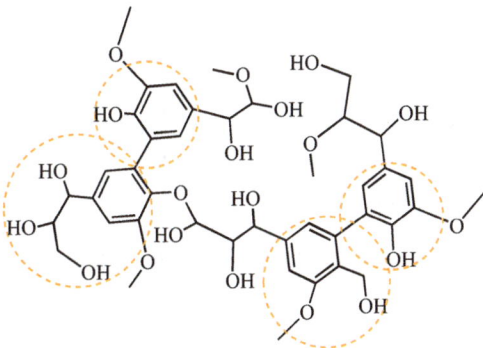

Figure 2.47: Simplified chemical structure of lignin.

2.7 Essential Keywords

Alditol The alcohol that results from the reduction of the aldehyde or keto group of an aldose or ketose.

Aldose A monosaccharide containing an aldehyde group.

Amylopectin A highly branched polysaccharide made up of about 2,000–200,000 glucose units.

Amylose A linear polysaccharide that is made up of about 500–20,000 glucose units.

Anomers Diastereomers that differ only in the configuration of the acetal or hemiacetal carbon of a sugar cyclic form.

Carbohydrates Polyhydroxyaldehydes and/or polyhydroxyketones.

Cellobiose A disaccharide with two units of glucose joined by β(1→4) glycosidic bond.

Chiral or stereogenic center A carbon atom that has four different groups bonded to it.

Disaccharides Glycosides formed from the linkage of two monosaccharides.

Epimerization A process in carbohydrate stereochemistry in which there is a change in the configuration of only one chiral center.

Furanose A five-membered ring system consisting of four carbon atoms.

Gentiobiose A disaccharide composed of two D-glucose units connected by a β(1→6) glycosidic bond.

Inulin A water-soluble polysaccharide consists of fructose units linked via β(2→1) linkages with a terminal glucose unit.

Ketose A monosaccharide containing a ketone group.

Koenigs-Knorr reaction The oldest glycosidation reaction that is effective for generating glucose β-glycosides.

Lignin A complex oxygen-containing macromolecule with an irregular phenolic structure.

Maltose A disaccharide made from two units of glucose.

Maltotetraose A tetrasaccharide made up of glucose molecules connected by α-1,4 glycosidic bonds.

Maltotriose A trisaccharide made up of three glucose units.

Monosaccharide Sugars with only one single unit.

Mutarotation The change in the equilibrium between two anomers due to optical rotation.

Oligosaccharides Saccharide molecules made up of 2–10 monosaccharide units formed through glycosidic bond polymerization.

Polysaccharide Monosaccharides joined together by glycosidic linkages.

Raffinose A trisaccharide that consists of a galactose unit coupled to sucrose.

Reducing sugars Sugars that contain hemiacetal and are in equilibrium with ring-opened form.

Trehalose A nonreducing natural disaccharide comprised of two glucose units connected together by α(1→1) glycosidic bond.

2.8 Practice Exercises

2.8.1   Why is **D**-glucose classified and named as aldohexose sugar?
2.8.2   List down the name of the main sugar units in raffinose.
2.8.3   What is the main difference between amylopectin and glycogen?
2.8.4   Why is sucrose considered a disaccharide? What are the sugar units in lactose?
2.8.5   Classify carbohydrates on the basis of sugar unit numbers.
2.8.6   Draw the chemical structure of any given ketohexose sugar and number the carbon atoms on the structure.
2.8.7   Give two examples for aldotetroses.
2.8.8   Show how β-**D**-glucopyranose is formed from **D**-glucose.
2.8.9   Draw the chemical structures of ribose and deoxyribose sugars.
2.8.10 What types of glycosidic linkages do sucrose, lactose, and maltose have?
2.8.11 What are the two enantiomers of glucose?
2.8.12 Draw a Fischer projection formula for the enantiomer of the following monosaccharide:

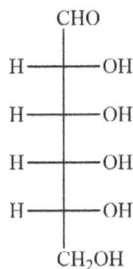

$$
\begin{array}{c}
\text{CHO} \\
\text{H} -\!\!\!-\!\!\!- \text{OH} \\
\text{H} -\!\!\!-\!\!\!- \text{OH} \\
\text{H} -\!\!\!-\!\!\!- \text{OH} \\
\text{H} -\!\!\!-\!\!\!- \text{OH} \\
\text{CH}_2\text{OH}
\end{array}
$$

2.8.13 Classify each of the following monosaccharides as **D**-isomer or **L**-isomer:

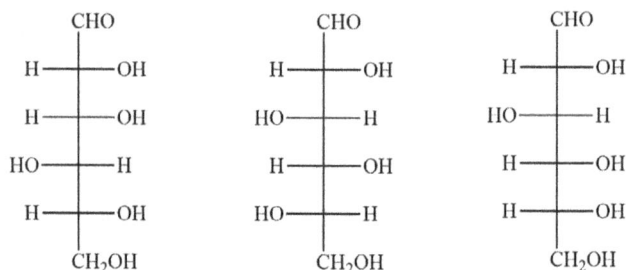

2.8.14 Classify each of the following monosaccharides according to both the number of carbon atoms and the type of carbonyl group present:

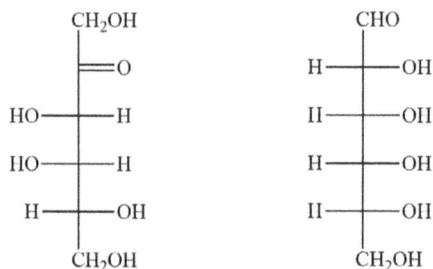

2.8.15 Indicate how many sugar units are present in each of the following:
   i.   Disaccharide
   ii. Oligosaccharide
   iii. Polysaccharide

2.8.16 Indicate whether or not each of the following is a correct characterization for the amylopectin form of starch:
   i.   It is a polysaccharide.
   ii. It contains two different types of monosaccharides.
   iii. It is a branched-chain glucose polymer.

2.8.17 Which monosaccharide is obtained from the hydrolysis of each of the following?
   i.   Sucrose
   ii. Glycogen
   iii. Starch

2.8.18 Classify each of the following polysaccharides as a glucose polymer or a glucose-derivative polymer:
  i. Chitin
  ii. Amylopectin
  iii. Glycogen

2.8.19 Name the sugar units in the following cellobiose structure:

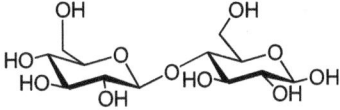

2.8.20 List names and abbreviations of three forms of cellulose.

****

Chapter 3
Lipids

3.1 Types (Classes) of Lipids

Lipids can be generally classified into waxes, triglycerides (triacylglycerols) phospholipids (PLs) (glycerophospholipids), sphingolipids, eicosanoids, and steroids (Figure 3.1).

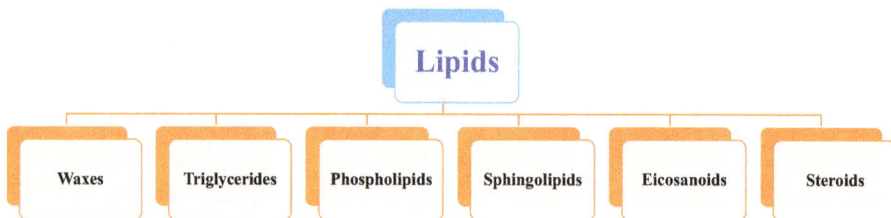

Lipids

| Waxes | Triglycerides | Phospholipids | Sphingolipids | Eicosanoids | Steroids |

Figure 3.1: Different classes of lipids.

3.2 Fatty Acids

Fatty acids (FAs) are long-chain carboxylic acids. They dissolve in fat solvents (hexane, ether, acetone, dichloromethane, dichloroethane, and chloroform) but not in water. Their chemical characteristics are defined by the carboxylic acid groups and hydrocarbon chains. The carboxylic acid group is polar and hydrophilic, whereas the hydrocarbon chain is nonpolar and hydrophobic. FAs that have only C–C single bonds are saturated. These are densely packed. They solidify at room temperature and have a high melting point. The second category is unsaturated FAs. They have low density and minimal interactions between chains because they include one or more C=C bonds in addition to C–C bonds. They are liquids at normal temperatures, with low melting points. Their general structures are depicted in Figure 3.2. Selected examples are also presented in Table 3.1 and Figure 3.3.

3.2.1 ω-3 and ω-6 Fatty Acids

The term "omega" refers to the position of the first double bond in the hydrocarbon chain, starting from the end farthest from the carboxyl group (also known as the methyl end or ω end). Omega-3 FAs form their initial double bond at the third carbon atom from the ω end. The chemical structure of alpha-linolenic acid (α-linolenic acid, ALA) is an 18-carbon chain with three double bonds on carbons 9, 12, and 15. The omega (ω) end of the chain is at carbon 18, while the double bond closest to the

https://doi.org/10.1515/9783111583273-003

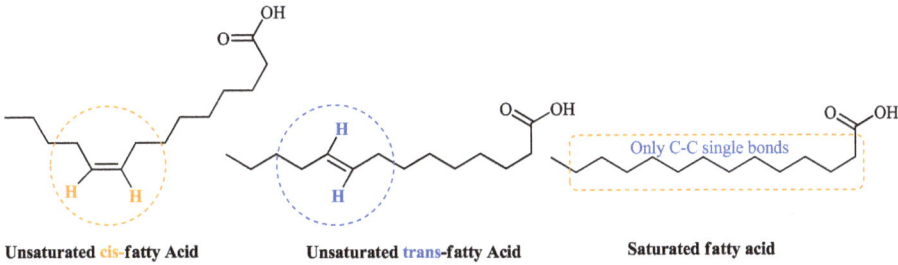

Unsaturated *cis*-fatty Acid   Unsaturated *trans*-fatty Acid   Saturated fatty acid

Figure 3.2: General chemical structures of fatty acids.

Table 3.1: Selected examples of fatty acids.

Fatty acid	Notation	Number of carbons	Formula	Type
Caproic	6:0	6	$CH_3(CH_2)_4$ COOH	**Saturated**
Caprylic	8:0	8	$CH_3(CH_2)_6$ COOH	
Capric	10:0	10	$CH_3(CH_2)_8$ COOH	
Lauric	12:0	12	$CH_3(CH_2)_{10}$ COOH	
Myristic	14:0	14	$CH_3(CH_2)_{12}$ COOH	
Palmitic	16:0	16	$CH_3(CH_2)_{14}$ COOH	
Stearic	18:0	18	$CH_3(CH_2)_{16}$ COOH	
Arachidic	20:0	20	$CH_3(CH_2)_{18}$ COOH	
Palmitoleic	16:1 (Δ^9) ω-7	16	$CH_3(CH_2)_5CH=CH(CH_2)_7$ COOH	**Mono Unsaturated**
Oleic	18:1 (Δ^9) ω-9	18	$CH_3(CH_2)_7CH=CH(CH_2)_7$ COOH	
Linoleic	18:2 $(\Delta^{9,\ 12})$ ω-6	18	$CH_3(CH_2)_4CH=CHCH_2CH=CH\ (CH_2)_7$ COOH	**Poly**
α-Linolenic	18:3 $(\Delta^{9,\ 12,\ 15})$ ω-3	18	$CH_3CH_2CH=CHCH_2CH=CH\ CH_2CH=CH$ $(CH_2)_7$ COOH	
γ-Linolenic	18:3 $(\Delta^{6,\ 9,\ 12})$ ω-6	18	$CH_3CH_2CH_2CH_2CH_2CH=CHCH_2CH=CH$ $CH_2CH=CH(CH_2)_4$ COOH	
Arachidonic	20:4 $(\Delta^{5,\ 8,\ 11,\ 14})$ ω-6	20	$CH_3(CH_2)_4\ (CH=CHCH_2)_4(CH_2)_2$ COOH	

omega carbon begins at carbon 15 = 18–3. Therefore, ALA is an ω-3 FA with ω = 18. Omega-6 FAs form their initial double bond at the sixth carbon atom from the ω end. The chemical structure gamma-linolenic acid (γ-linolenic acid, GLA) is an 18-carbon chain with three double bonds on carbons 6, 9, and 12. The omega (ω) end of the

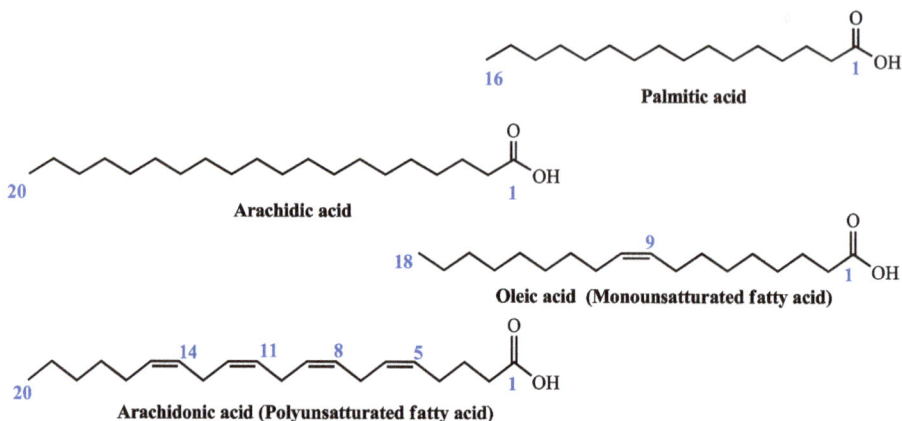

Figure 3.3: Chemical structures of selected fatty acids.

chain is at carbon 18, while the double bond closest to the omega carbon begins at carbon 12 = 18–6. Therefore, GLA is an ω-6 FA with ω = 18 (Figure 3.4).

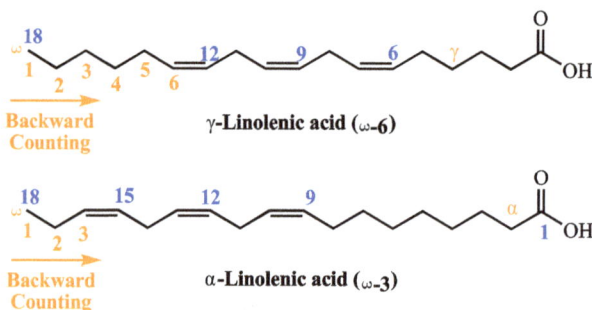

Figure 3.4: The chemical structures of α-linolenic acid (ALA) and the γ-linolenic acid (GLA).

3.2.2 Unsaturated Fatty Acids with Triple Bonds

There are a couple of triple-bond FAs. These are 17-octadecyonic acid (17-ODYA) and 5,8,11,14-eicosatetraynoic acid (ETYA). Octadec-17-ynoic acid is an acetylenic FA that is octadecanoi acid (stearic acid) which has been doubly dehydrogenated at positions 17 and 18 to give the corresponding alkynoic acid. It is a long-chain FA, an acetylenic FA, a terminal acetylenic compound, and a monounsaturated FA. Eicosa-5,8,11,14-tetraynoic acid is a long-chain FA. A 20-carbon unsaturated FA contains four alkyne bonds (Figure 3.5).

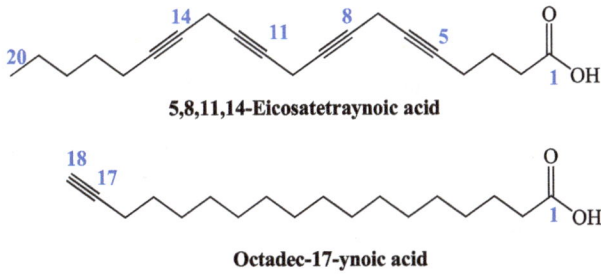

5,8,11,14-Eicosatetraynoic acid

Octadec-17-ynoic acid

Figure 3.5: The chemical structures of 17-octadecyonic acid (17-ODYA) and 5,8,11,14-eicosatetraynoic acid (ETYA).

3.3 Waxes

They are chemical compounds composed of long aliphatic alkyl chains of FA esters such as beeswax from honeycomb and carnauba wax derived from Brazilian palm tree leaves. Waxes can also contain a variety of functional groups including primary and secondary alcohols, ketones, aldehydes, and aromatic rings, as seen in montan wax derived from geological sources such as mineral deposits, as well as synthetic waxes such as microcrystalline wax, which has a more complex structure and smaller crystal sizes than natural wax. Figure 3.6 displays some of these waxes.

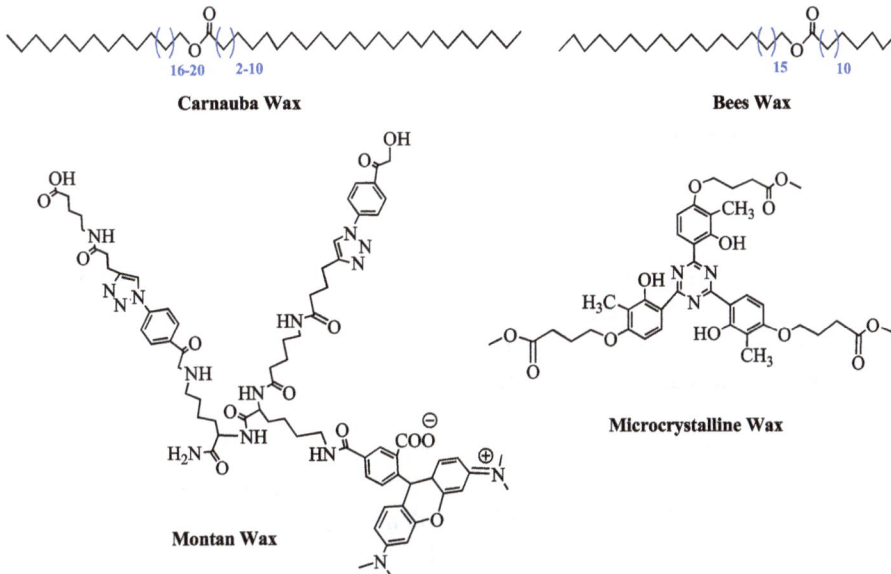

Carnauba Wax

Bees Wax

Montan Wax

Microcrystalline Wax

Figure 3.6: The chemical structures of beeswax, carnauba wax, montan wax, and microcrystalline wax.

3.4 Fats and Oils (Triglycerides or Triacylglycerols)

Triglycerides (TGs) are glycerol (1,2,3-propanetriol) triesters. They are also known chemically as triacylglycerols. The three FAs in any given triacylglycerol are not necessarily the same. Fats and oils are mostly composed of mixed TGs containing different FAs. The fats and solids are primarily composed of saturated FAs. The melting point of TGs lowers as the number of double bonds increases. In oils, for example, the main TGs are those containing unsaturated FAs. The type of TG depends on its origin. Saturated TGs are the most common in mammals, but unsaturated TGs are more prevalent in plants.

The primary distinction between unsaturated and saturated TGs is that unsaturated TGs have double bonds between the carbon atoms, whereas saturated TGs have just one single bond between the carbon atoms. TGs are created by esterifying glycerol with three FAs, as demonstrated in Figure 3.7.

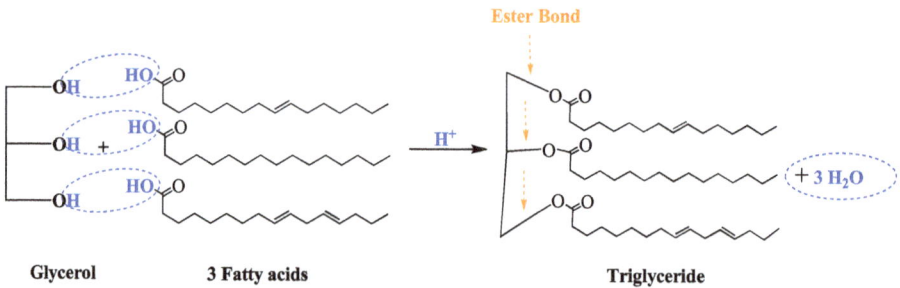

Figure 3.7: Triglyceride formation.

3.4.1 Naming Triglycerides

TGs can be named as follows:
- Count the carbons on the glycerol units 1, 2, and 3 from the top down. Each carbon in the glycerol molecule is numbered using the "stereospecific numbering (sn) system" as sn-1, sn-2, and sn-3.
- Identify each FA name linked in the TG.
- Identify the position of the FAs and replace "ic" or "eic" with "oyl."

The chemical structure of the TG shown in the first example in Figure 3.8A consists of a saturated at sn-1, monounsaturated at sn-2 and polyunsaturated FA at sn-3. Its proper name is *1-palmitoyl 2-oleoyl 3-linolenoyl glycerol* or in shorthand POL (where P is palmitic acid, O is oleic acid, and L is linolenic acid).

In the second example, the chemical structure of the TG shown in Figure 3.8B consists of a saturated at sn-1, monounsaturated at sn-2, and saturated FA at sn-3. Its proper name is *1-lauroyl-2-oleoyl-3-palmitoyl glycerol* or in shorthand LOP (where

L is lauric acid, O is oleic acid, and P is palmitic acid). In the third example, the chemical structure of the TG shown in Figure 3.8C consists of a polyunsaturated FA at *sn*-1, monounsaturated at *sn*-2, and saturated FA at *sn*-3. Its proper name is ***1-linoleoyl-2-oleoyl-3-stearoyl glycerol*** or in shorthand LOS (where L is linoleic acid, O is oleic acid, and S is stearic acid). In the fourth example, the chemical structure of the TG shown in Figure 3.8D consists of a saturated FA at *sn*-1, polyunsaturated at *sn*-2, and saturated FA at *sn*-3. Its proper name is ***1,3-distearoyl-2-linoleoyl glycerol*** or in shorthand SLS (where S is stearic acid, L is Linoleic acid, and S is stearic acid).

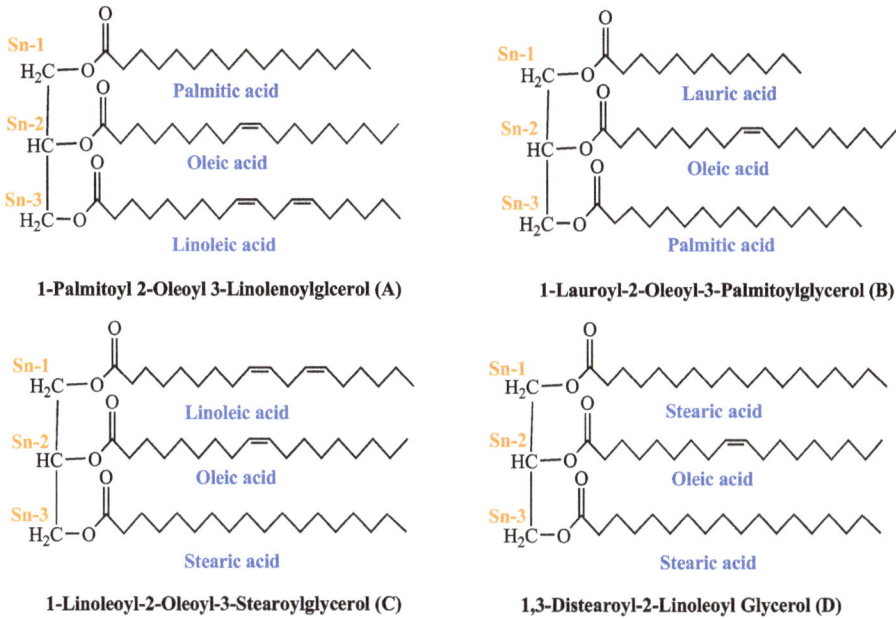

1-Palmitoyl 2-Oleoyl 3-Linolenoylglcerol (A)

1-Lauroyl-2-Oleoyl-3-Palmitoylglycerol (B)

1-Linoleoyl-2-Oleoyl-3-Stearoylglycerol (C)

1,3-Distearoyl-2-Linoleoyl Glycerol (D)

Figure 3.8: Selected examples of triglycerides and their names.

3.4.2 Chemical Reactions of Triglycerides

Hydrogenation: unsaturated TGs react with hydrogen (H_2) in the presence of nickel (Ni) or platinum (Pt) catalysts to form saturated TGs. The C=C bonds are converted to C–C bonds (Figure 3.9).

Hydrolysis: TGs split by water and acid or enzyme catalyst to glycerol and three FAs. They also split into glycerol and the salts of FAs "soaps" by the action of strong base such as KOH or NaOH in a reaction called saponification as shown in Figure 3.10.

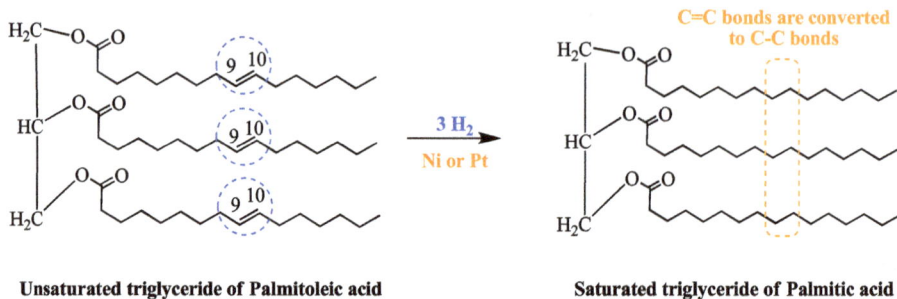

Figure 3.9: Hydrogenation of unsaturated triglycerides.

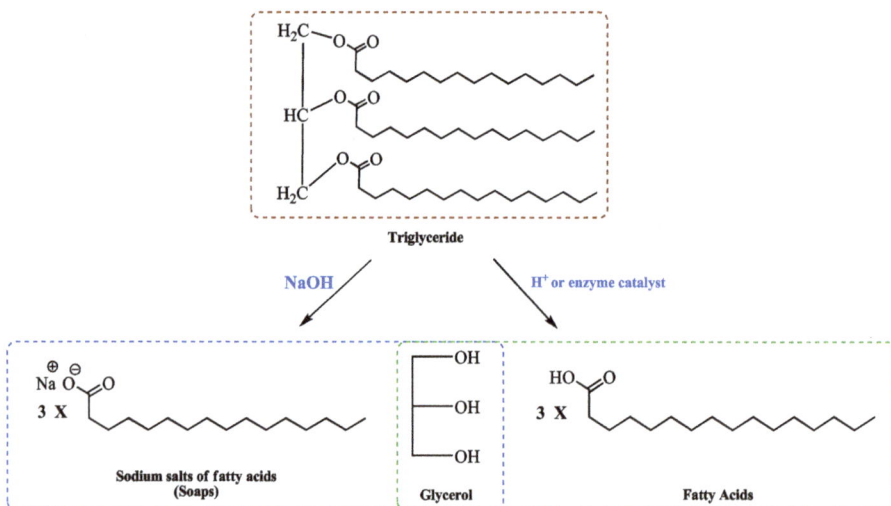

Figure 3.10: Hydrolysis of triglycerides.

3.5 Phospholipids

Phosphatidic acid (Figure 3.11) is a molecule formed by esterifying glycerol with two molecules of either of the three FAs (palmitic acid, stearic acid, or oleic acid) and one of phosphoric acid.

PLs or glycerophospholipids are a class of polar lipids composed of two FAs, a glycerol unit, and a phosphate group that is esterified to an amino alcohol derivative such as 2-aminoethanol, choline, and serine or inositol to produce various PLs. Lecithins comprise phosphoric acid, choline, and two FAs (Figure 3.12A). Cephalins are divided into two types: ethanol amines ("phosphatidyl ethanolamines") (Figure 3.12B) and serine ("phosphatidyl serine") (Figure 3.12C). Phosphoinositides, also known as inositol phosphatides, have a glycerol backbone esterified to two FA chains and phosphate linked to

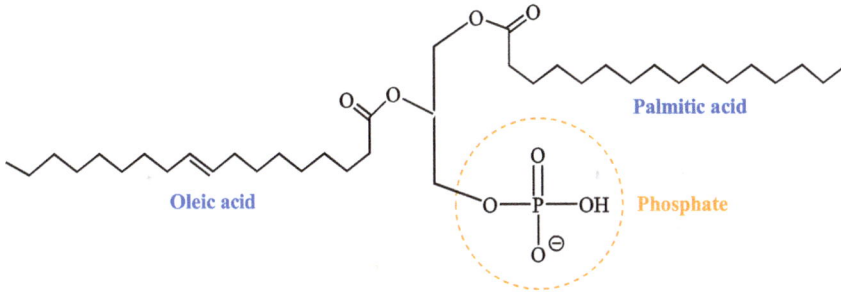

Figure 3.11: Phosphatidic acid.

a polar head group myoinositol (Figure 3.12D). Plasmalogens are glycerophospholipids with a vinyl-ether linkage at *sn*-1 and an ester linkage at *sn*-2 (Figure 3.12E).

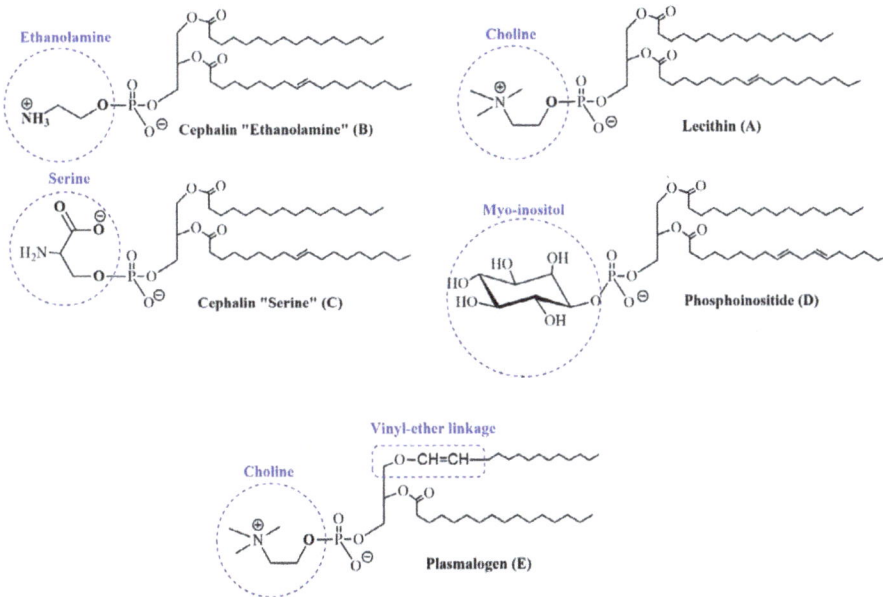

Figure 3.12: Examples of phospholipids.

3.6 Sphingolipids

Sphingolipids are lipids that contain sphingosine, an organic aliphatic amino alcohol (Figure 3.13A). Ceramides are among the simplest sphingolipids (Figure 3.13B). Glycosphingolipids (Figure 3.13C) are broad sphingolipids containing one or more sugar molecules (glucose or galactose). The only sphingolipids that contain phosphorus are phosphos-

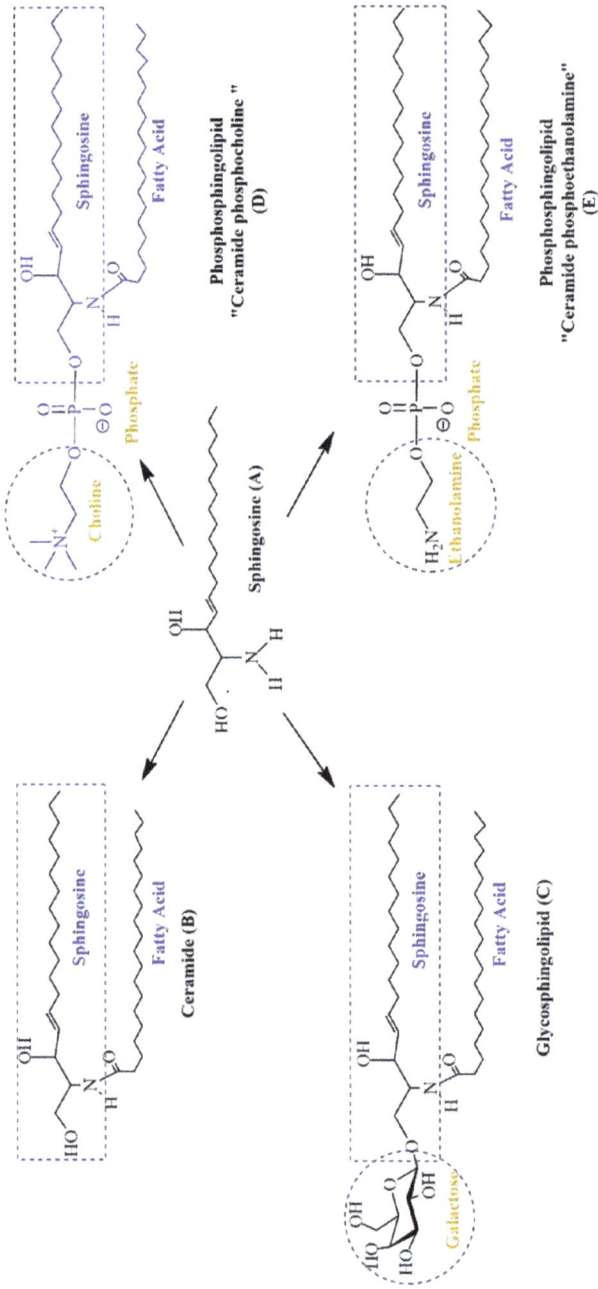

Figure 3.13: Types of sphingolipids.

phingolipids, also known as sphingomyelins. Phosphodiester bonds hold them together. Ceramide phosphocholines are the most common phosphosphingolipids found in mammals (Figure 3.13D). Ceramide phosphoethanolamines are present in insects (Figure 3.13E).

3.7 Eicosanoids

Eicosanoids are a class of compounds formed from 20-carbon polyunsaturated FAs, most often arachidonic acid (the Greek word "eicosa" means "20"). Figure 3.14 lists examples of eicosanoids such as thromboxanes, leukotrienes, prostaglandins, and lipoxins.

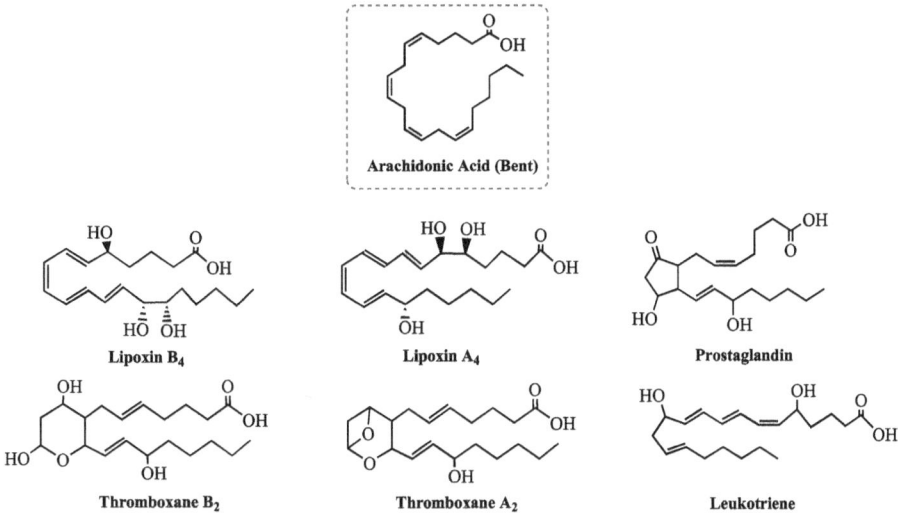

Figure 3.14: Selected chemical structures of eicosanoids and arachidonic acid.

3.8 Steroids

The core, or nucleus, of a steroid, is known as gonane or sterane (cyclopentanoperhydrophenanthrene). Typically, it is made up of 17 carbon atoms fused into four fused rings: the D ring, which is a five-membered cyclopentane ring, and the rings A, B, and C are three six-membered cyclohexane rings. Diagrams of plane projections are used to depict steroid structures. Note that the methyl groups (C18 and C19) swap hydrogens for C10 and C13. Other frequently occurring substituents include side chains at C17, similar to the structure of cholesterol, and hydroxyl groups at C3 (Figure 3.15). All of the sex hormones, bile acids, sterols, adrenal cortex hormones, molting hormones of insects, and numerous other physiologically active compounds of plants and animals belong to the steroid group.

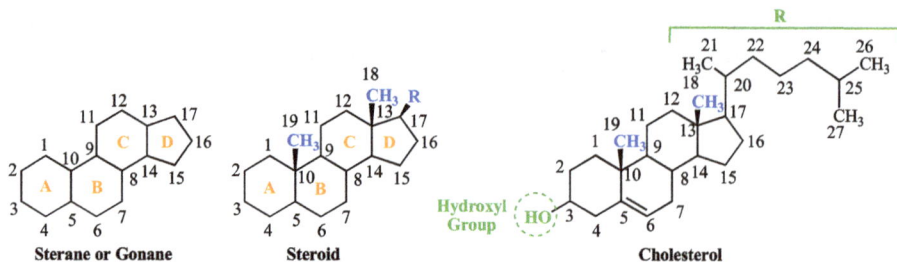

Figure 3.15: Chemical structures of gonane (sterane), steroid, and cholesterol.

To specify the stereochemistry of rings A and B, indicate whether the hydrogen atom connected at C5 is above or below the diagram's plane (β or α). The α-, β-sign represents the orientation of a substituent group bonded to a saturated (completely substituted) carbon in the steroid ring structure. The bonding of β-attached substituents is shown by a solid line, while that of α-substituents is shown by a broken line. If the solid line connecting the -H group to the ring is cis to the methyl group, and both are above the page plane, the notation is 5β-H. In the 5α-H, the line is dashed and oriented trans to the methyl group. Each carbon atom in a steroid molecule is assigned a number, which is designated for a specific place in the hypothetical parent skeletal structure, regardless of whether a carbon atom occupies that position (Figure 3.16).

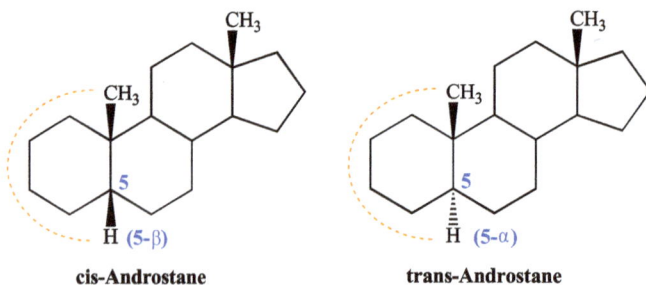

Figure 3.16: The α-, β-symbolism in steroids.

3.8.1 Sex Hormones

Sex hormones are chemicals that regulate reproduction and sexual desire. Common female sex hormones include estrogen and progesterone, whereas testosterone is prevalent in most males (Figure 3.17).

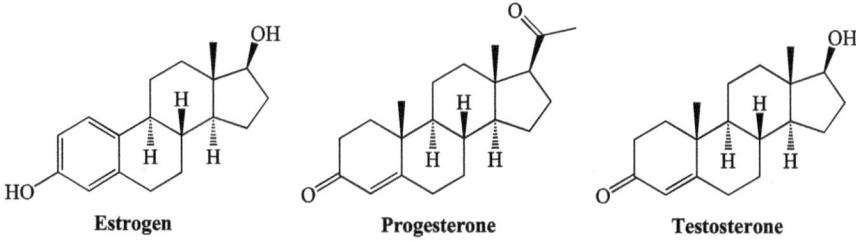

Figure 3.17: Chemical structures of sex hormones.

3.8.2 Adrenal Cortex Hormones (Corticosteroids)

The most important among these types are aldosterone, cortisol, and cortisone. The kidneys use aldosterone to regulate salt levels in the blood and tissues. Cortisol stimulates the metabolism and consumption of carbohydrates, proteins, and fats. Cortisol is a naturally occurring corticosteroid metabolite that functions as a pharmaceutical precursor (Figure 3.18).

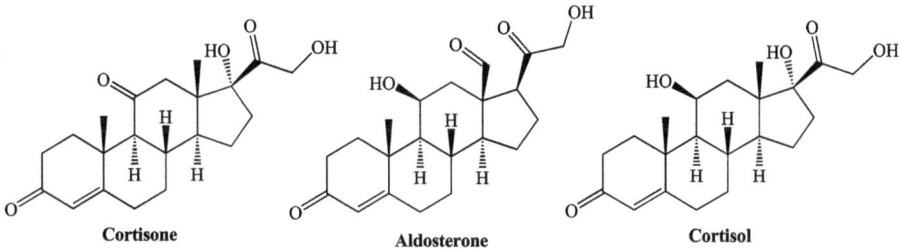

Figure 3.18: Chemical structures of adrenal cortex hormones.

3.8.3 Sterols

Sterols (3β-hydroxysteroids) are hydroxylated derivatives of gonane. Sterols have a complicated ring structure that includes a hydroxyl group at C-3 and a flexible side chain with 8–10 carbons at C-17. The best-known sterol is cholesterol. Cholesterol (also known as cholesteryl alcohol or cholest-5-en-3β-ol) is a derivative of cholestane. It has the chemical formula $C_{27}H_{45}OH$ and consists of three regions: a hydrocarbon tail, a ring structure region with four hydrocarbon rings, and a hydroxyl group (Figure 3.19).

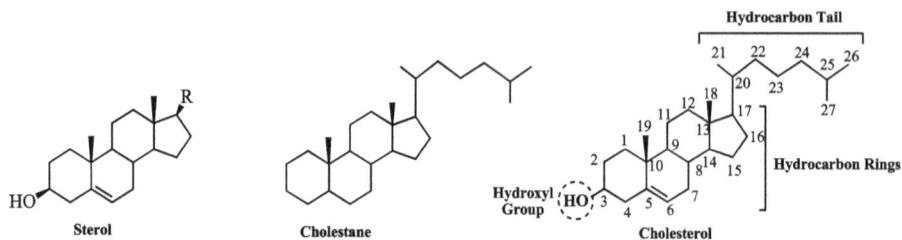

Figure 3.19: Chemical structures of sterol, cholestane, and cholesterol.

3.8.4 Bile Acids

Bile acids are a broad family of steroidal compounds produced in the liver from cholesterol and actively released into the bile alongside cholesterol and PLs. The principal bile acids are cholic and chenodeoxycholic acid, and the secondary bile acids are deoxycholic acid and lithocholic acid (Figure 3.20).

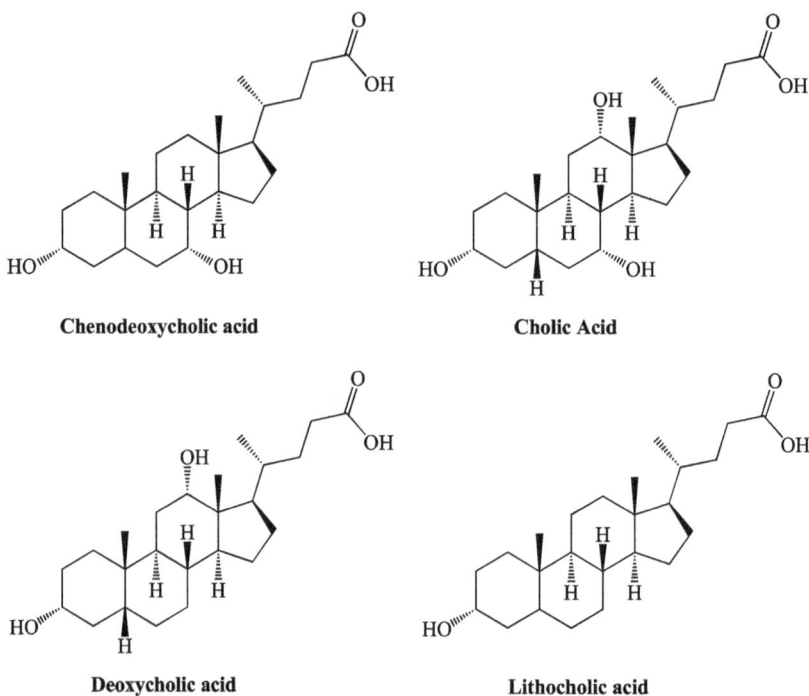

Figure 3.20: Chemical structures of bile acids.

3.8.5 Steroid Molting Hormones

Ecdysone, also known as α- or β-ecdysone, and 20-hydroxyecdysone are closely related steroid molting hormones (Figure 3.21).

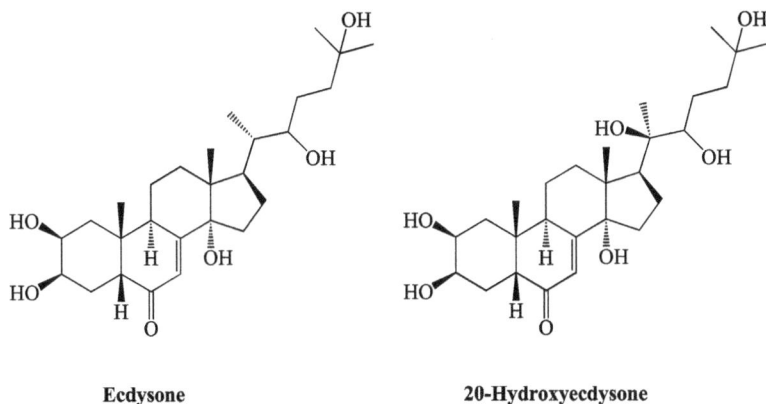

Ecdysone   **20-Hydroxyecdysone**

Figure 3.21: Chemical structures of steroid molting hormones.

3.9 Terpenes

Terpenes typically consist of two, three, four, or six isoprene units (isoprene is 2-
methyl-1,3-butadiene, $CH_2=C(CH_3)-CH=CH_2$). Terpenes can be cyclic or acyclic, with a
wide range of structural options. Figure 3.22 shows a few examples of them. Terpenes
are also classified according to the number of isoprene units and their structure, with
monoterpenes (C10), sesquiterpenes (C15), diterpenes (C20), and triterpenes (C30).

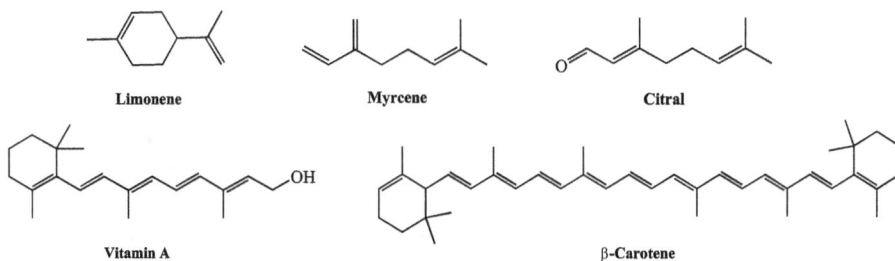

Limonene   **Myrcene**   **Citral**

Vitamin A   **β-Carotene**

Figure 3.22: Chemical structures of selected terpenes.

3.10 Terpenoids

Terpenoids are branching lipids, also known as isoprenoids, that occur when the isoprene molecule is chemically modified (Figure 3.23). These lipids have many physiological roles in plants and animals as well as practical uses such as medicines (capsaicin), pigments (β-carotene and xanthophyll), and fragrances (menthol, camphor, limonene, and pinene).

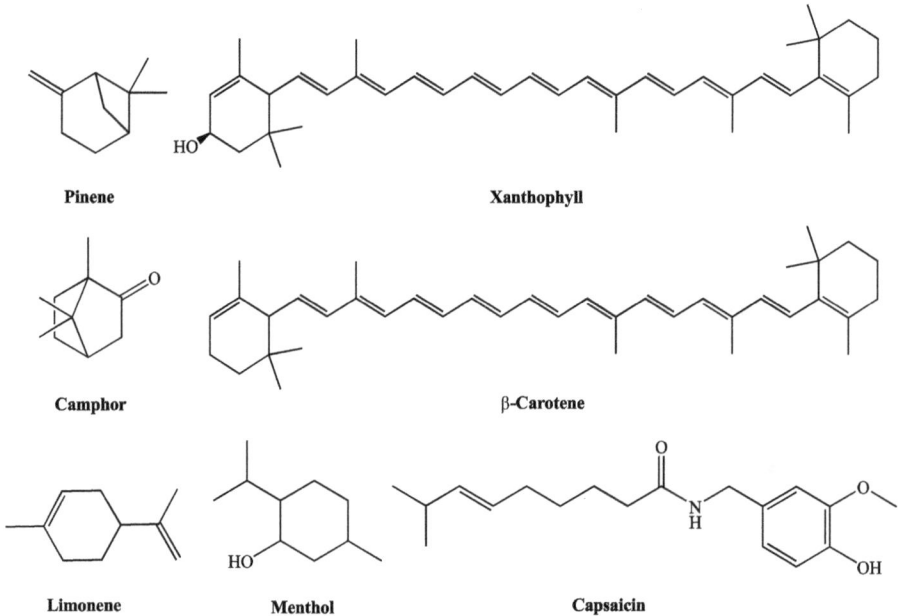

Figure 3.23: Chemical structures of selected terpenoids.

3.11 Vitamins

A vitamin is an organic molecule that plays key roles in the human body and must be received through dietary sources in our meals. Vitamins are categorized into two broad types based on their solubility: water-soluble vitamins and fat-soluble vitamins. Figure 3.24 depicts nine water-soluble vitamins (thiamin "vitamin B_1," riboflavin "vitamin B_2," niacin "vitamin B_3," pantothenic acid "vitamin B_5," pyridoxine "vitamin B_6," cyanocobalamin "vitamin B_{12}," ascorbic acid "vitamin C," folic acid or folate "vitamin 9," and biotin "vitamin B_7"). Figure 3.25 depicts four fat-soluble vitamins (vitamin A exists in three forms: retinol, retinal, and retinoic acid, vitamin D in two forms: ergocalciferol and cholecalciferol, vitamin E in two forms: tocopherol and tocotrienol, and vitamin K in two forms: phylloquinone and menaquinone).

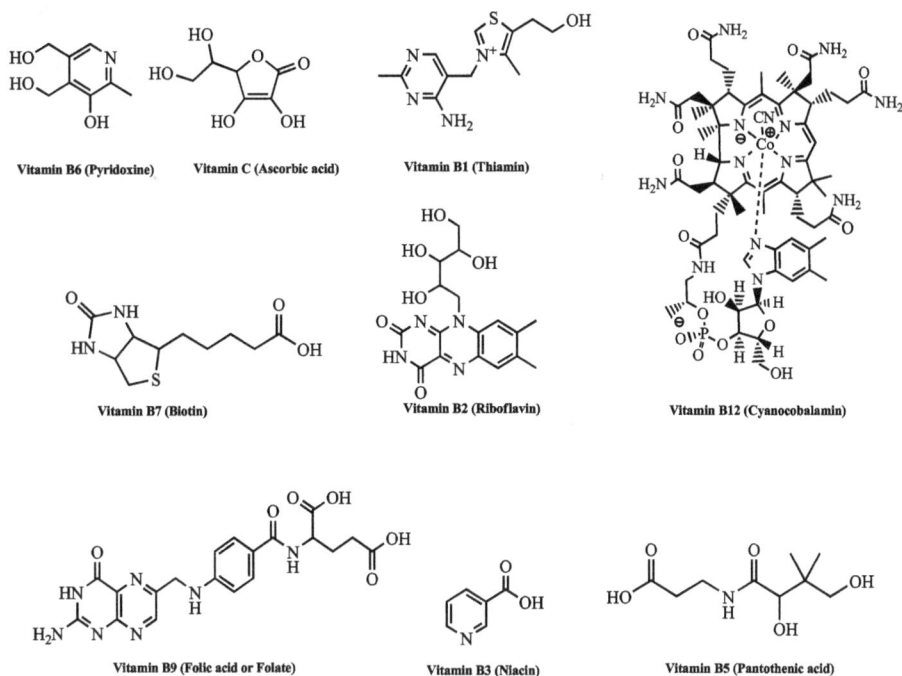

Figure 3.24: Chemical structures of selected water-soluble vitamins.

3.12 Conjugated Lipids or Lipid Polymer Conjugates

Conjugated lipids are lipid molecules that have covalent bonds with additional functional groups. These functional groups contribute to the biological system's diversity of functions. One notable example is *lipid-protein conjugation*, a new strategy for creating innovative delivery systems that combine proteins with lipids. Furthermore, these conjugates outperform single carriers regarding synergistic action and desired features within the body. Another example is the cholesterol conjugate of oligonucleotides. Cholesterol is a lipophilic compound that may successfully bind to LDL and HDL. The *cholesterol conjugate of oligonucleotides* can help them pass through lipid membranes and enter cells and tissues. (Chol = cholesterol; ASO = antisense oligonucleotides; LDL = low-density lipoprotein; HDL = high-density lipoprotein). The third example is *PEGylated lipids*. They are made up of linear polyethylene glycol (PEG) chemically attached to the polar head of a lipid. PEG is a hydrophilic, flexible, and inert polymer. It's chemical structure is 1,2-distearoyl-*sn*-glycero-3-phosphoethanolamine-*N*[methoxy(PEG)]. Examples of these conjugated lipids are shown in Figure 3.26.

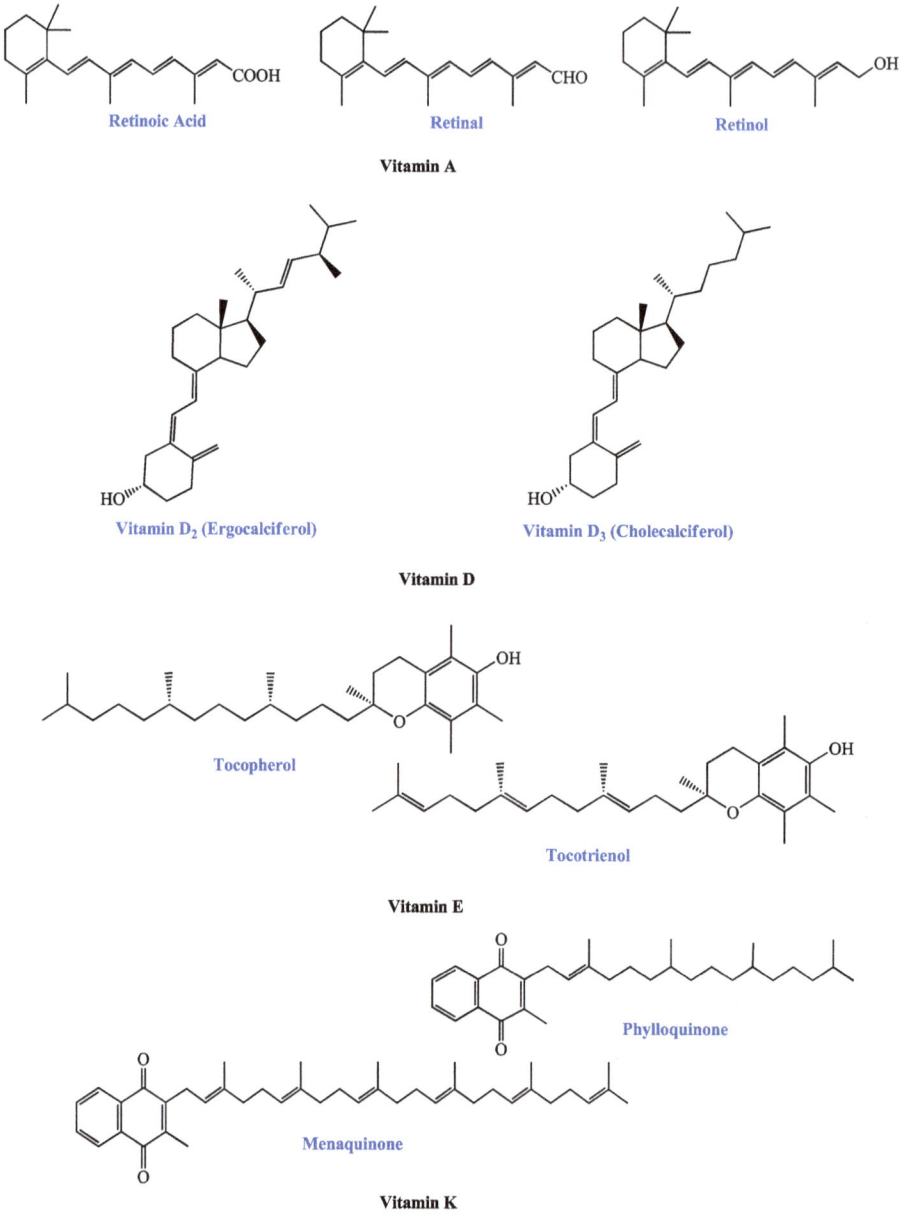

Retinoic Acid   **Retinal**   **Retinol**

Vitamin A

Vitamin D₂ (Ergocalciferol)   **Vitamin D₃ (Cholecalciferol)**

Vitamin D

Tocopherol

Tocotrienol

Vitamin E

Phylloquinone

Menaquinone

Vitamin K

Figure 3.25: Chemical structures of selected fat-soluble vitamins.

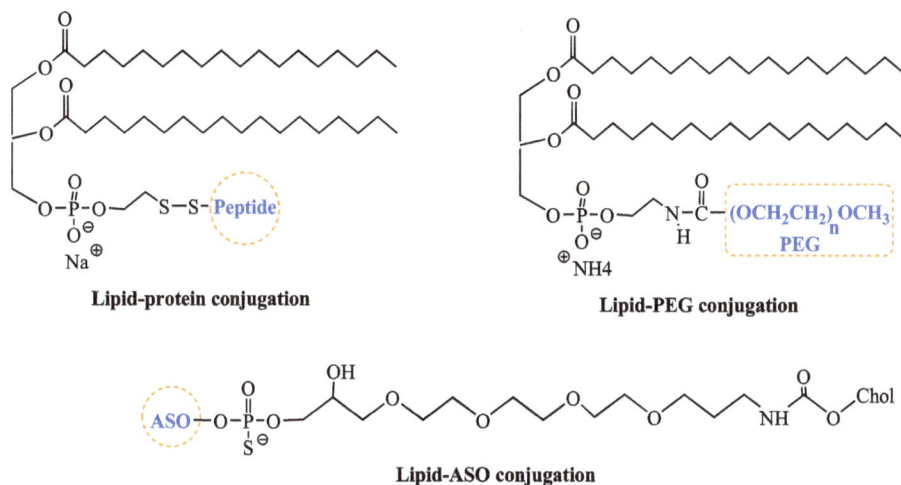

Figure 3.26: Selected general examples of the lipid polymer Conjugates.

3.13 Essential Keywords

Bile acids A broad family of steroidal compounds produced in the liver from choles-terol and actively released into the bile alongside cholesterol and PLs.

Cephalins Phosphoglycerides with ethanolamine or serine esterified to the phospho-ric acid group.

Conjugated lipids Lipid molecules that have covalent bonds with additional func-tional groups can be peptides or oligonucleotides.

Eicosanoids A class of compounds formed from 20-carbon polyunsaturated FAs, most often arachidonic acid.

Fatty acids Long-chain carboxylic acids with even numbers of carbon atoms between 12 and 26.

Fat solvents Solvents that are able to dissolve fats such as hexane, ether, acetone, dichloromethane, dichloroethane, and chloroform.

Glycerides FA esters of glycerol.

Lecithins Phosphoglycerides with choline esterified to the phosphoric acid group.

Lipids Substances that are not soluble in water and can be extracted from tissues by nonpolar organic solvents called fat solvents.

Omega Refers to the last methyl carbon in the FA and also the position of the first double bond in the hydrocarbon chain, starting from the end farthest from the carboxyl group.

Phosphatidic acid A molecule in which glycerol is esterified with two molecules of FAs and one phosphoric acid.

Phospholipids Lipids derived from phosphatidic acid.

Phosphoglyceride An ester of glycerol in which the three hydroxy groups are esterified by two FAs and one phosphoric acid derivative.

Saponification Base-promoted hydrolysis of an ester to make soap.

Saturated fatty acids Long-chain carboxylic acids with only carbon-carbon single bonds.

Sex hormones Chemicals that regulate reproduction and sexual desire.

Soap The sodium or potassium salts of FAs.

Sphingolipids Lipids that contain sphingosine, an organic aliphatic amino alcohol.

Sterols (3β-hydroxysteroids) Hydroxylated derivatives of gonane.

Steroids A fused tetracyclic ring system that involve three six-membered rings and one five-membered ring.

Terpenes A family of compounds with carbon skeletons composed of two or more five-carbon isoprene units.

Terpenoids Branching lipids, also known as isoprenoids, occur when the isoprene molecule is chemically modified.

Triglyceride Glycerol (1,2,3-propanetriol) triesters.

Unsaturated fatty acids Long-chain carboxylic acids with one or more C=C bonds in addition to C–C bonds.

Vitamins Organic molecules play key roles in the human body and must be received through dietary sources in our meals.

Wax An ester of long-chain FA with long-chain alcohol.

3.14 Practice Exercises

3.14.1 Match the following statements in column **A** with the correct answer in column **B** provided in the table:

Column A	Column B
1. The notation for palmitic acid is_____	A. Long-chain carboxylic acids
2. Fatty acids are _____	B. Hydrogenation catalysts
3. The structures of steroids are based on _____	C. 20-Carbon carboxylic acids
4. Eicosanoids are a group of _____	D. Fused tetracyclic ring system
5. Nickel (Ni) and platinum (Pt) are _____	E. 16:0

3.14.2 Which one of the following fatty acid structures has the **cis** configuration? Why?

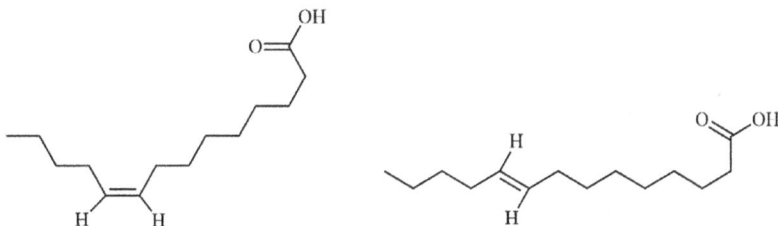

3.14.3 What does the following notation 16:1 (Δ^9) ω-7 mean?

3.14.4 Complete the missing information (**?**) in the following hydrogenation reaction:

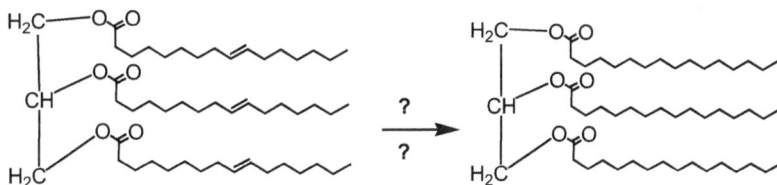

3.14.5 Assign numbers to the atoms of the following steroid structure:

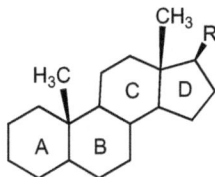

3.14.6 Draw the chemical structure of oleic acid and show the location of the double bond.

3.14.7 Write down the shorthand notations of the following fatty acids structure:

i.

ii.

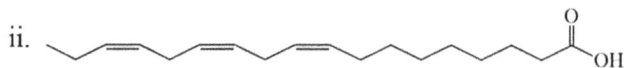

iii.

3.14.8 Show and name the three regions in the following cholesterol structure:

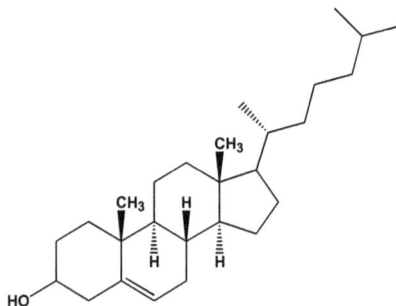

3.14.9 Draw the chemical structure of carnauba wax.

3.14.10 What group of lipids are esters of long-chain fatty acids and long-chain alcohols?

3.14.11 How many fatty acids are required to produce one molecule of a fat or oil?

3.14.12 What are the two functional groups that chemically react to form TG?

3.14.13 Ω-6 (ω-6) and Ω-3 (ω-3) fatty acids are important in the area of nutrition. Which of the following structures would be classified as an Ω-3 (ω-3) fatty acid?

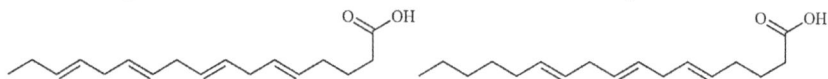

3.14.14 Which of the following fatty acids is omega-3 fatty acid?

    i.   Oleic acid

    ii. Linolenic acid

    iii. Arachidonic acid

3.14.15 Which of the following terms best describes the structure below?

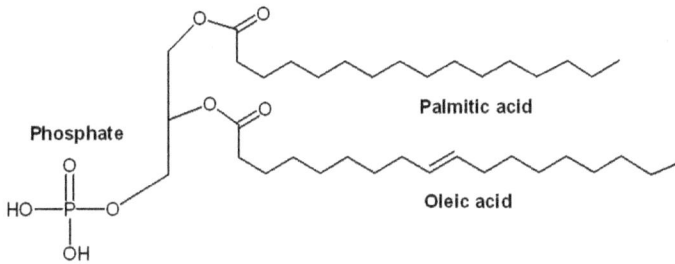

Phosphate

Palmitic acid

Oleic acid

3.14.16 Label the hydrophilic and hydrophobic regions of the molecule shown below.

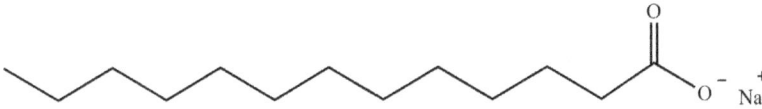

3.14.17 Of the two vitamins A and C, one is hydrophilic and water-soluble while the other is hydrophobic and fat-soluble. Which is which?

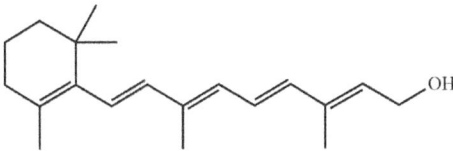

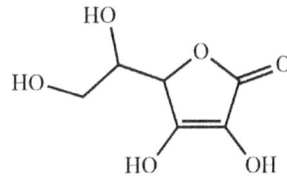

Vitamin A   Vitamin C

3.14.18 Which of the following is a second name for phosphatidylcholine?
    i.   Steroid
    ii. Lecithin
    iii. Cephalin

3.14.19 Name four terpenoids that are used in fragrances.

3.14.20 Give four examples of bile acids.

****

Chapter 4
Proteins (Polymers of Amino Acids)

4.1 Definition

Proteins are polymers of amino acids. They can be formed from several polypeptide chains or single chains. Nonpeptide components, such as saccharide chains and lipids, may also be in their structure. Peptide bonds hold amino acid residues together in proteins. Figure 4.1 depicts all levels of protein structure.

4.2 Amino Acids (Proteins Building Blocks)

The amino acid structure has four distinct groups (except for glycine, which has three). The amino group is $-NH_2$, the carboxyl group is $-COOH$, the hydrogen is H, and the side chain is R (Figure 4.2). Proteins including amino acids with the amino group linked to the carbon atom close to the carboxyl group are referred to as alpha- or 2-amino acids.

4.2.1 Structures and Nomenclature of Amino Acids

The type of side chain "R" in an amino acid's structure sets it apart from others. It is not always a simple hydrocarbon group, such as alanine, or a single hydrogen atom, like glycine. Other active groups found on the side chain are hydroxyl, thiol, amino, and carboxyl. Amino acids are classed according to the properties of their side chains. The nine nonpolar amino acids that have hydrophobic side chains are glycine (Gly), alanine (Ala), valine (Val), leucine (Leu), isoleucine (Ile), proline (Pro), phenylalanine (Phe), methionine (Met), and tryptophan (Trp). Six amino acids include polar but uncharged side chains. These include serine (Ser), threonine (Thr), cysteine (Cys), asparagine (Asn), glutamine (Gln), and tyrosine (Tyr). Aspartic acid (Asp) and glutamic acid (Glu) are acidic amino acids, while arginine (Arg), lysine (Lys), and histidine (His) are basic amino acids (Figure 4.3).

The 20 amino acids comprise the building blocks for the synthesis of proteins. The remaining two additional amino acids are derived by modification after the biosynthesis of the protein. Hydroxyproline and cystine are synthesized from proline and cysteine, respectively, after the protein chain has been synthesized. Hydroxyproline is produced by hydroxylation of proline by the enzyme prolyl hydroxylase. Cysteine is oxidized under mild conditions to disulfide cystine. The disulfide linkage is important in maintaining the overall shape of a protein (Figure 4.4).

https://doi.org/10.1515/9783111583273-004

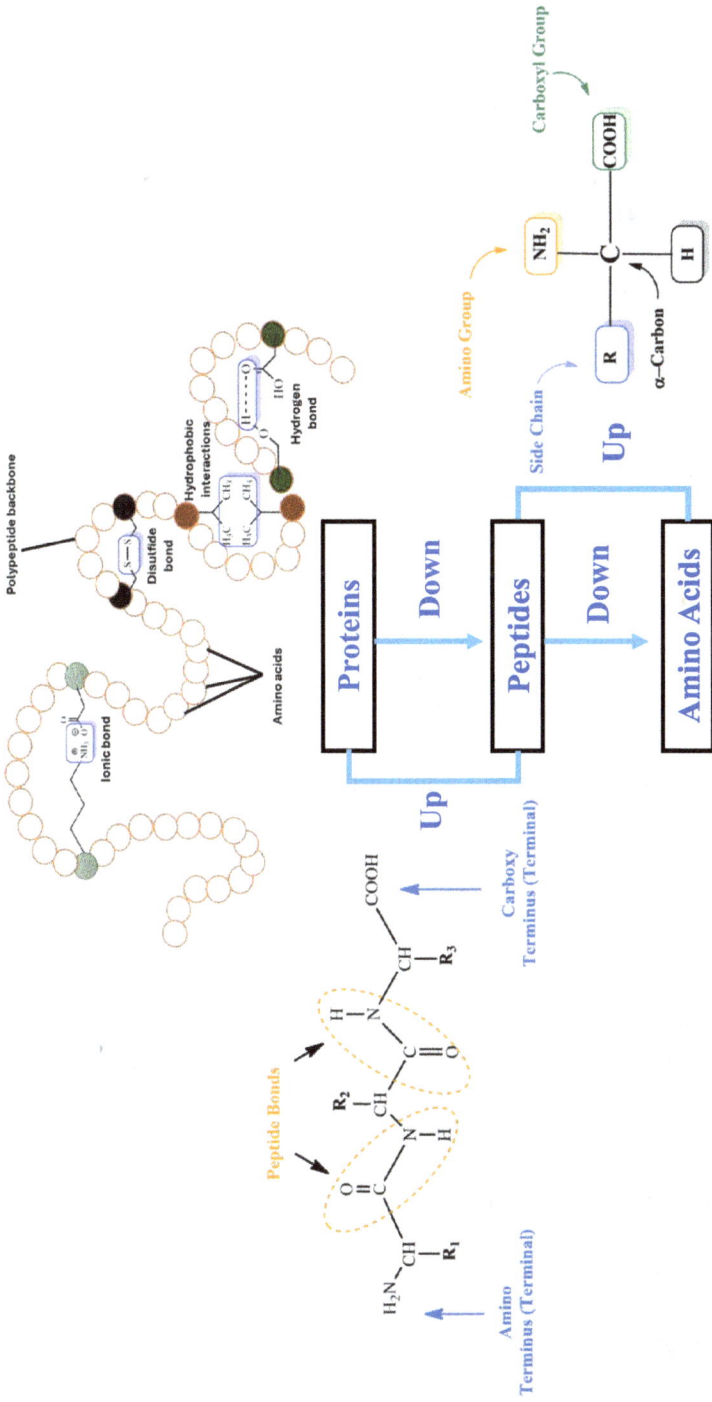

Figure 4.1: Levels of protein structure.

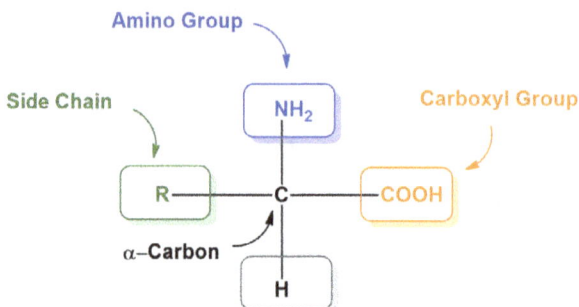

Figure 4.2: General structure of amino acids.

4.2.2 Synthesis of α-Amino Acids

Direct ammonolysis of α-halo acid results in low yields. This is a straightforward nucleophilic substitution in which ammonia combines with an a-halo carboxylic acid. Figure 4.5 illustrates the synthesis of alanine from α-bromopropionic acid.

Another way is to synthesize amino acids from potassium phthalimides. This is a high-yielding variation of the Gabriel synthesis. Figure 4.6 depicts an excellent method for preparing glycine from potassium phthalimide.

The Strecker synthesis involves treating an aldehyde with ammonia and hydrogen cyanide to produce an α-aminonitrile, which is then hydrolyzed to produce the α-amino acid. The reaction proceeds through an intermediate imine. Figure 4.7 illustrates a good method for synthesizing alanine from acetaldehyde.

4.2.3 Resolution of D,L-Amino Acids

A racemic amino acid mixture can be resolved by converting D,L-amino acids to a racemic mixture of N-acylamino acids, which is then hydrolyzed with a deacylase enzyme to selectively deacylate the L-acylamino acid (Figure 4.8).

4.2.4 Acid-Base Behavior of Amino Acids

The amino acid molecule has both a basic amino group and an acidic carboxyl group. Amino acids can exist in the zwitterion form by internally transferring a hydrogen ion from the $-COOH$ group to the $-NH_2$ group, resulting in an ion with both negative and positive charge.

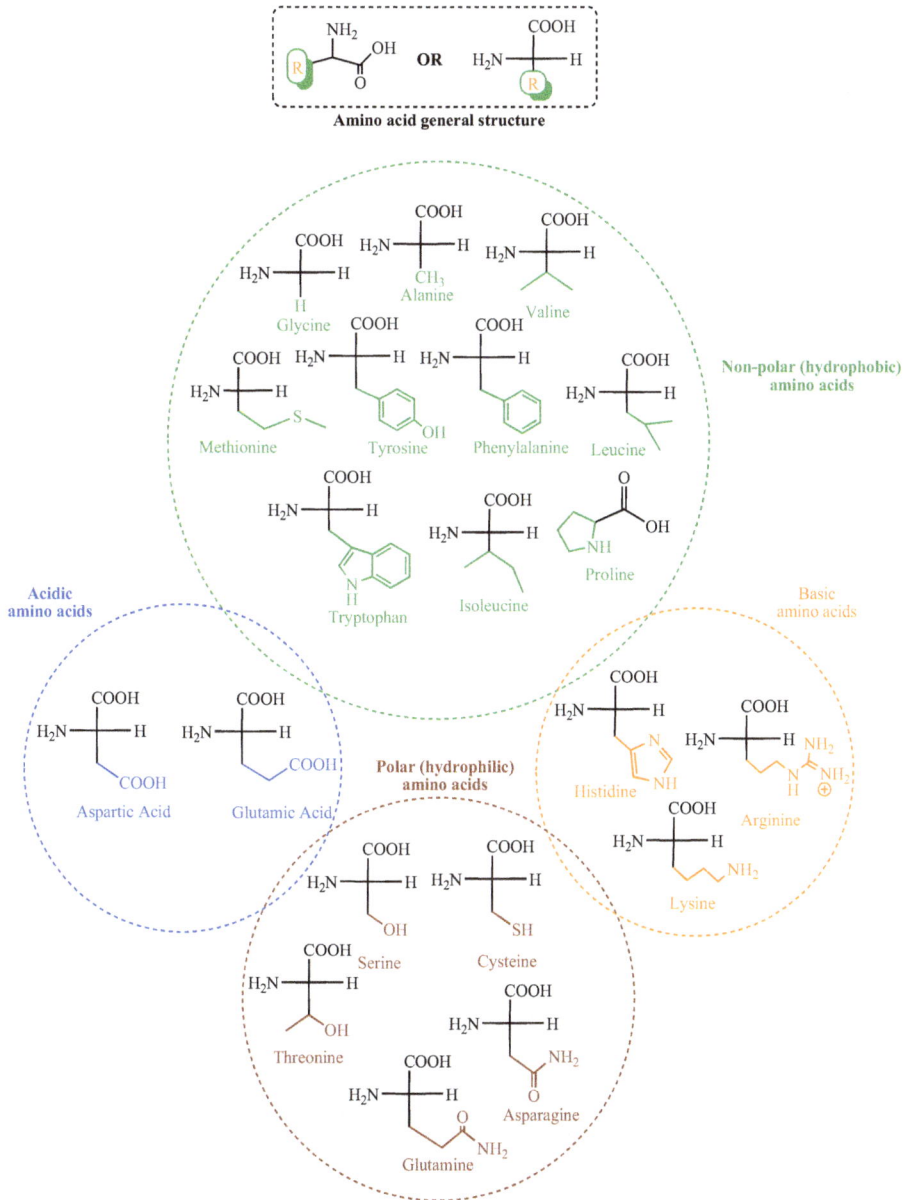

Figure 4.3: Selected amino acid structures.

In the dry solid form, amino acids exist as dipole ions (zwitterions). In an aqueous solution, there is an equilibrium between the dipolar ion and the cationic-anionic forms of amino acids. The prevailing form is determined by the pH of the solution. At low pH, the amino acid is largely cationic, with a net charge of +1. At high pH, the

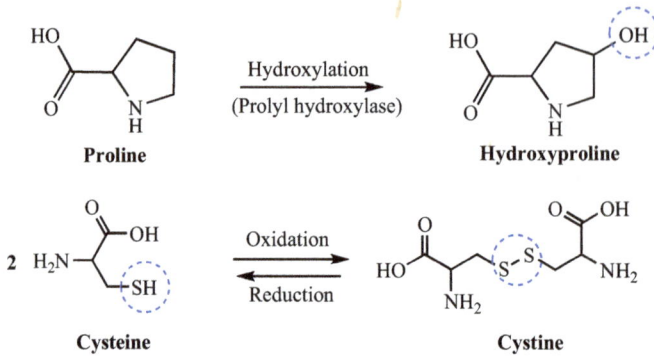

Figure 4.4: Hydroxylation of hydroxyproline and oxidation of cysteine.

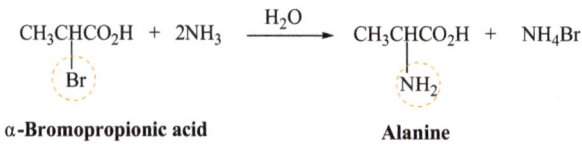

Figure 4.5: Synthesis of alanine from α-bromopropionic acid.

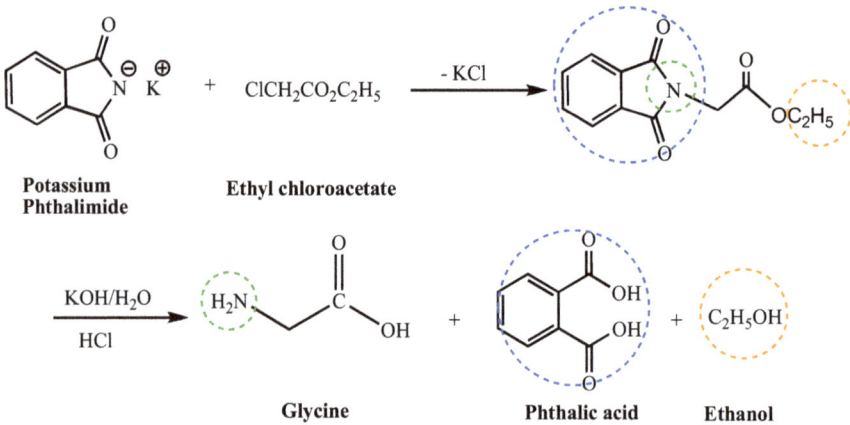

Figure 4.6: Synthesis of glycine from potassium phthalimide.

amino acid exists primarily in anionic form, with a net charge of −1. At the isoelectric point (pI or IEP, the pH at which an amino acid has no net electrical charge), the dipolar ion concentration is maximum, the concentrations of anionic and cationic forms are equal, and the net charge is zero (Figure 4.9).

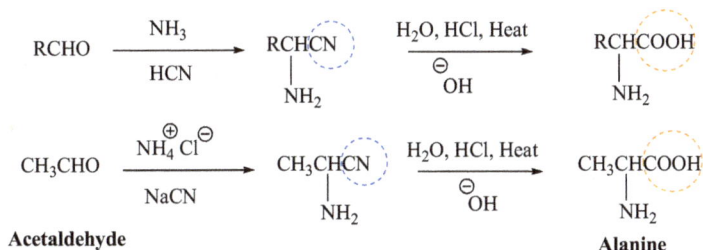

Figure 4.7: The Strecker synthesis.

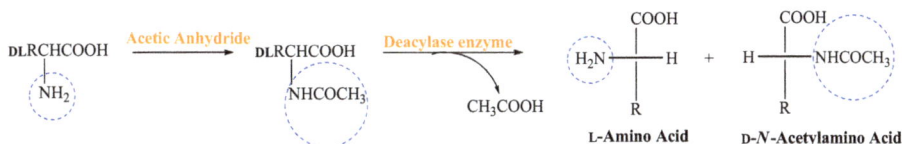

Figure 4.8: Resolution of D,L-amino acids.

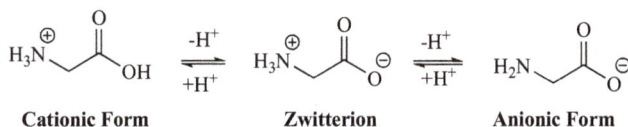

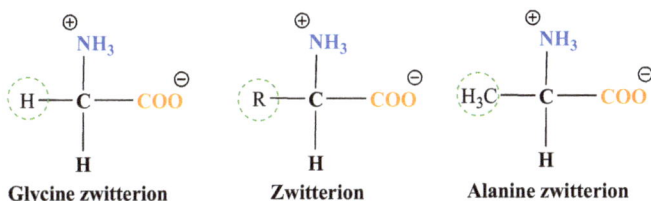

Figure 4.9: Acid-base behavior of amino acids with examples.

4.2.5 Separation of Amino Acids

Amino acids can be separated and purified using electrophoresis. This method is based on the movement of charged particles in an electric field. A combination of amino acids is deposited in the center of a sheet of cellulose acetate. The sheet is soaked in an aqueous buffered solution with a pH of 6. At this pH, aspartic acid exists as a −1 ion, alanine as zwitterions, and lysine as a +1 ion. An electric current causes the negatively charged amino acid (aspartic acid) to migrate to the (+) electrode while

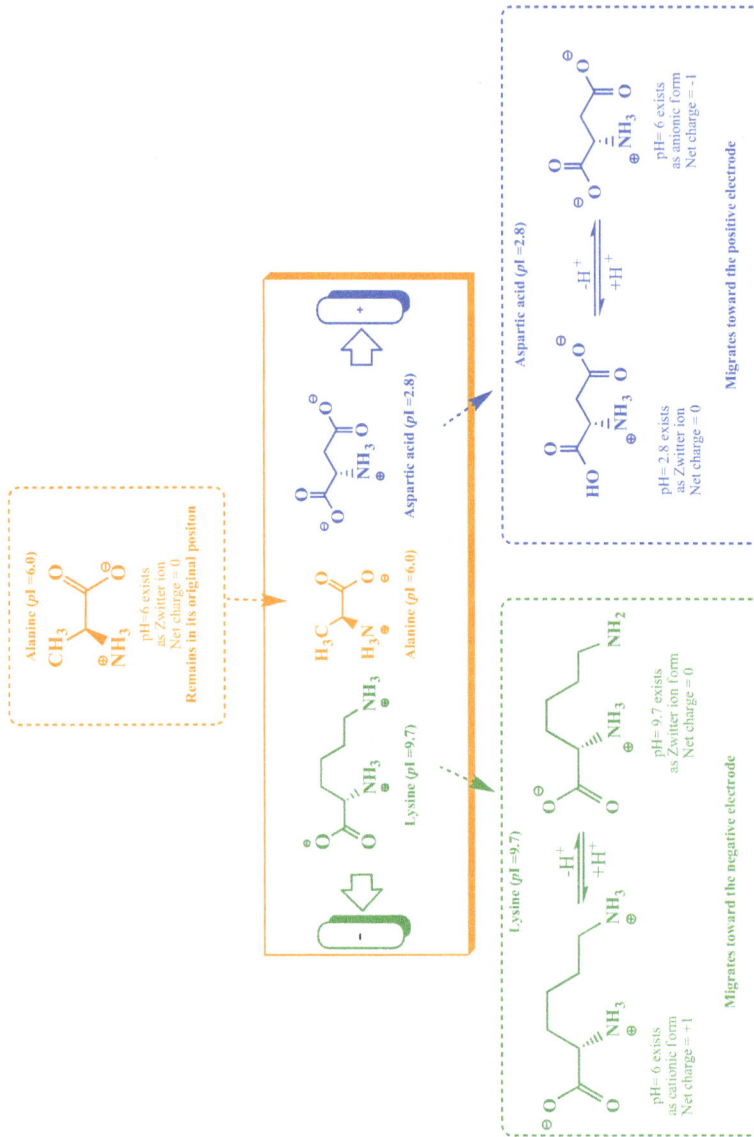

Figure 4.10: The basics of amino acid separation by electrophoresis.

the positively charged one (lysine) migrates to the (–) electrode. The amino acid (alanine), which exists as zwitterions and has a net charge of zero, remains in its original position (Figure 4.10).

4.2.6 Chirality of Amino Acids

In all chiral amino acids, the stereogenic carbon in the structure's chiral center is bonded to four different groups. Amino acids can occur in two forms: L and D. The "R" group in glycine is another hydrogen atom, which is why it is achiral ("not chiral") (Figure 4.11).

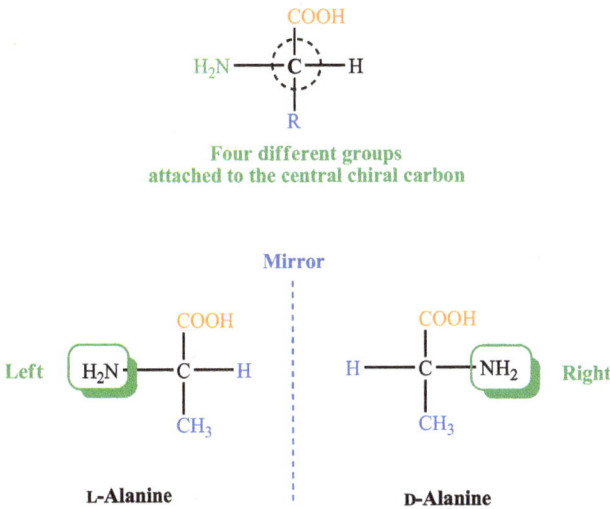

Figure 4.11: Chirality of amino acids.

4.3 Peptide Bond

When the –COOH group of one amino acid molecule combines with the –NH_2 group of another amino acid molecule, a water molecule "H_2O" is released and a peptide or amide bond "CO–NH" is created (Figure 4.12). This type of reaction is called the dehydration reaction (also known as the condensation reaction).

4.4 Polypeptides (Polypeptide Chains)

Amino acids combine to create unbranched amino acid chains known as peptides, and when the number of amino acids in the chain is considerable, it is referred to as

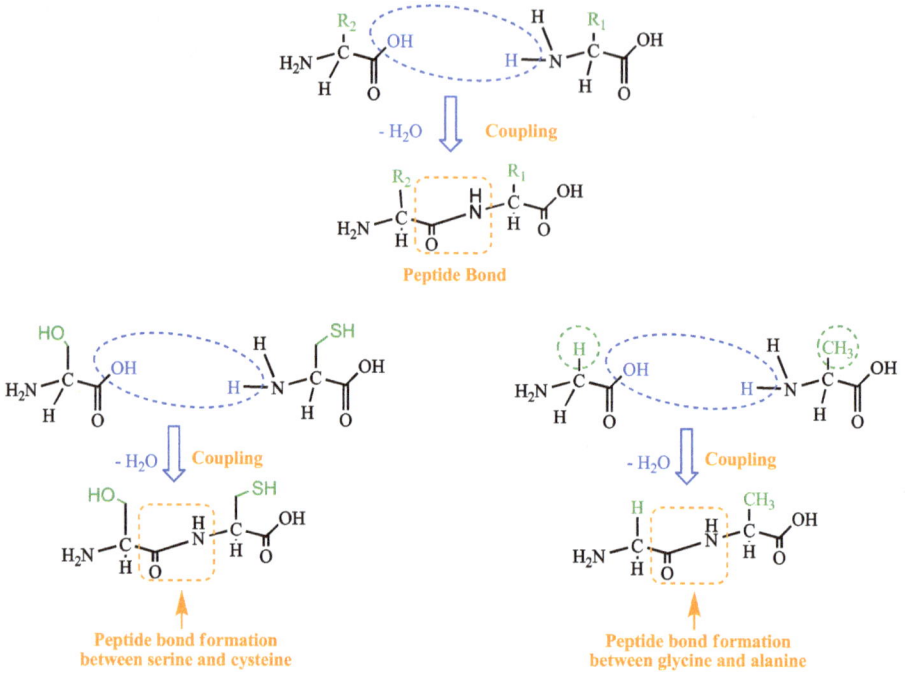

Figure 4.12: Peptide bond formation with examples.

a polypeptide chain (Figure 4.13). Each peptide chain can be several or many amino acids long. They are further classed based on the number of amino acids found in the chain. A peptide with two amino acids is known as a dipeptide, three amino acids as a tripeptide, and so on. The number of Rs in the chain denotes the number of amino acids, while Rs-1 signifies the number of peptide bonds in the same chain.

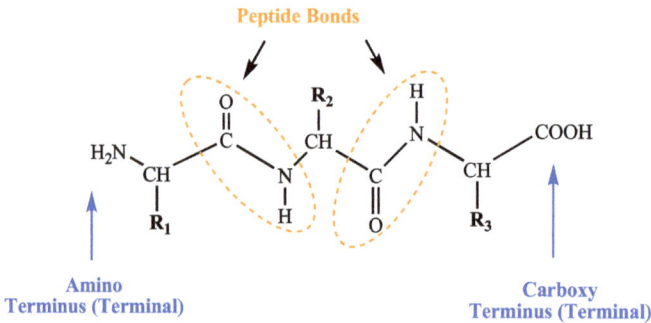

Figure 4.13: Tripeptide chain.

4.5 Peptide Nomenclature

The following IUPAC rules can be applied to name small peptides:
- **Rule 1**: The C-terminal amino acid residue (located at the far right of the structure) maintains its full amino acid name.
- **Rule 2**: All other amino acid residues have names that end in -yl. The -yl suffix replaces the -ine or -ic acid ending of the amino acid name, except for tryptophan (tryptophyl), cysteine (cysteinyl), glutamine (glutaminyl), and asparagine (asparaginyl).
- **Rule 3**: The amino acid naming sequence begins at the N-terminal amino acid residue.

For example, the peptide chain Glu-Ser-Ala has three amino acids: Glu, serine, and alanine. Alanine, the C-terminal residue (on the far right), retains its entire name. The remaining amino acid residues in the peptide have "shortened" designations that finish in -yl. The -yl suffix replaces the -ine or -ic acid endings of amino acid names. As a result, glutamic acid is converted to glutamyl, serine to seryl, while alanine remains unchanged. The IUPAC name, which specifies the amino acids in the sequence from N-terminal to C-terminal residue, is changed to glutamylserylalanine.

The second peptide chain, Gly-Tyr-Gln-Asn, has four amino acids: glycine, tryptophan, glutamine, and asparagine. The IUPAC name, which specifies the amino acids in the sequence from N to C, is changed to glycyltryptophylglutaminylasparagine. The third peptide chain, Gly-Tyr-Ser-Ser, has the IUPAC name glycyltyrosylserylserine. Figure 4.14 summarizes the three examples.

Glu–Ser–Ala

This peptide chain has three amino acids: glutamic acid, serine, and alanine. The IUPAC name, which lists the amino acids in the sequence from N-terminal residue to C-terminal residue, becomes glutamylserylalanine.

Gly–Trp–Gln–Asn

In this peptide chain, the four amino acids present are glycine, tryptophan, glutamine, and asparagine. The IUPAC name, which lists the amino acids in the sequence from N-terminal residue to C-terminal residue, becomes glycyltryptophyl glutaminyl asparagine.

Gly–Tyr–Ser–Ser

This peptide chain also has four amino acids, namely glycine, tyrosine, serine, and serine. The IUPAC name is glycyltyrosylserylserine.

Figure 4.14: Naming of selected peptides.

4.6 Peptide Synthesis (The General Principles)

During peptide synthesis, amino acids' functional groups must be protected so that they do not interfere with the formation of the peptide bond. Peptide bond formation involves four steps. These are protection, activation, coupling, and selective deprotection or full deprotection, as illustrated in Figure 4.15.

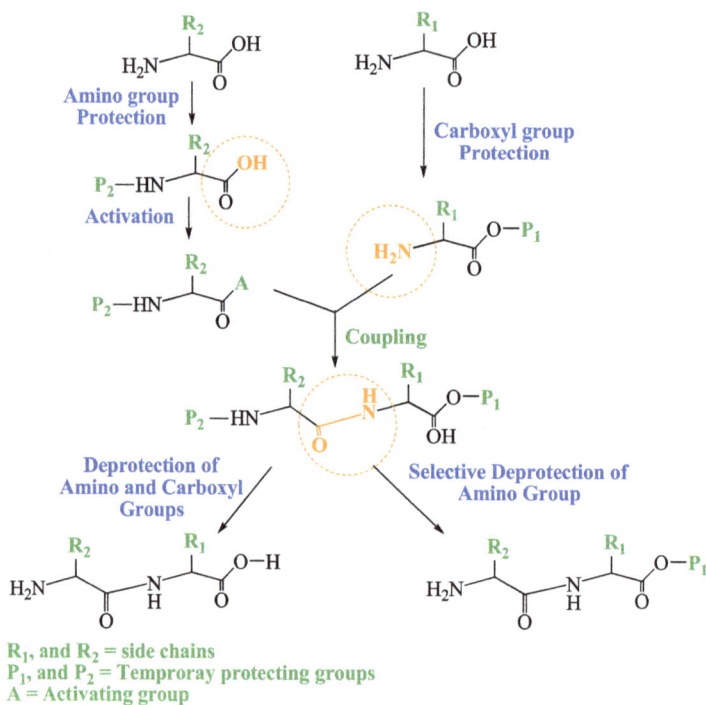

R_1, and R_2 = side chains
P_1, and P_2 = Temproray protecting groups
A = Activating group

Figure 4.15: General principles of peptide synthesis.

4.6.1 Dipeptide Synthesis

The four steps required to synthesize a dipeptide such as Ala-gly are depicted in Figure 4.16. The butyloxycarbonyl (BOC) derivative protects alanine's amino group, whereas the methyl ester protects glycine's carboxyl group. The two protected amino acids were then combined with dicyclohexylcarbodiimide (DCC). Following synthesis, the acid treatment removes the BOC-protecting group, while basic hydrolysis removes the methyl ester.

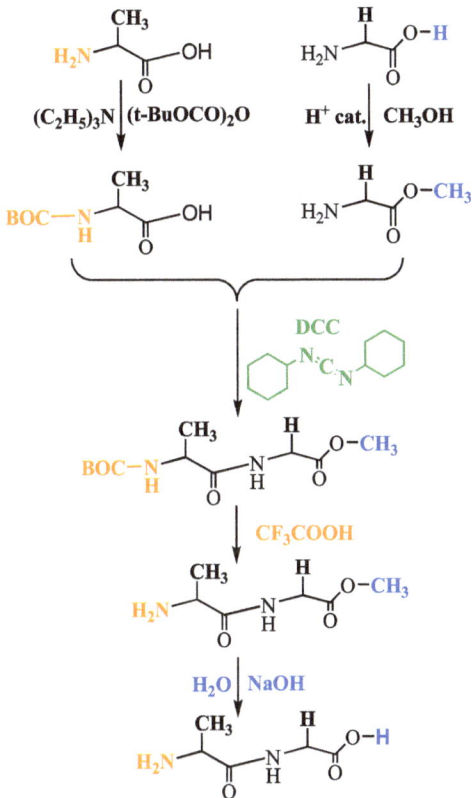

Figure 4.16: Dipeptide synthesis.

4.6.2 Solid-Phase Synthesis for Peptides and Polypeptides

Solid-phase peptide synthesis is an automated process for producing conjugated pepti-des and normal polypeptides. The synthesis is based on the sequential assembly of a peptide chain connected to a solid support as illustrated in Figure 4.17.

4.7 Levels of Protein Structure

4.7.1 Primary Structure

Each protein's basic structure determines its unique folding pattern. The basic struc-ture of a protein (polypeptide) chain is the linear sequence of amino acids linked to-gether by peptide bonds (Figure 4.18).

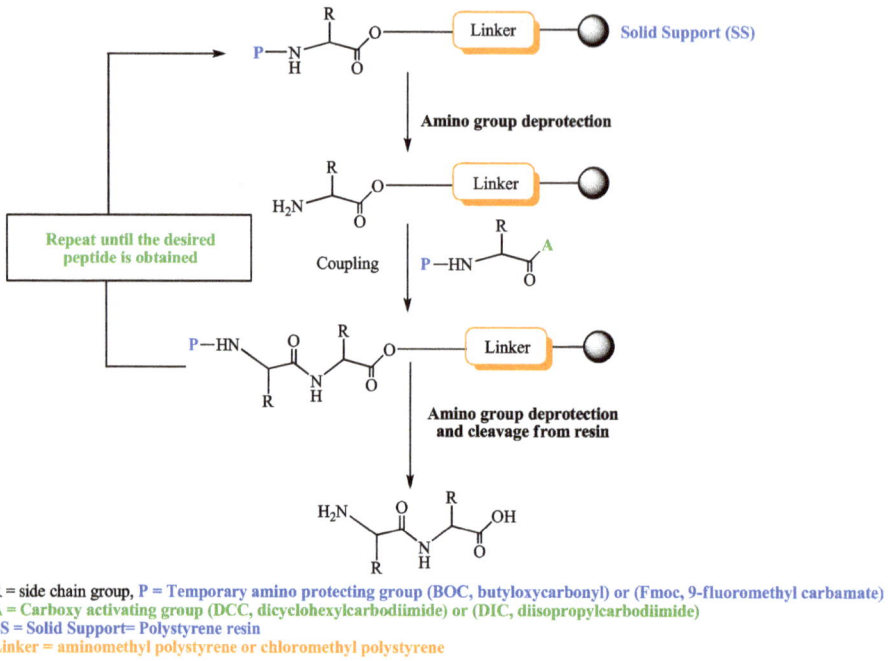

R = side chain group, P = Temporary amino protecting group (BOC, butyloxycarbonyl) or (Fmoc, 9-fluoromethyl carbamate)
A = Carboxy activating group (DCC, dicyclohexylcarbodiimide) or (DIC, diisopropylcarbodiimide)
SS = Solid Support= Polystyrene resin
Linker = aminomethyl polystyrene or chloromethyl polystyrene

Figure 4.17: Peptide and polypeptide solid-phase synthesis.

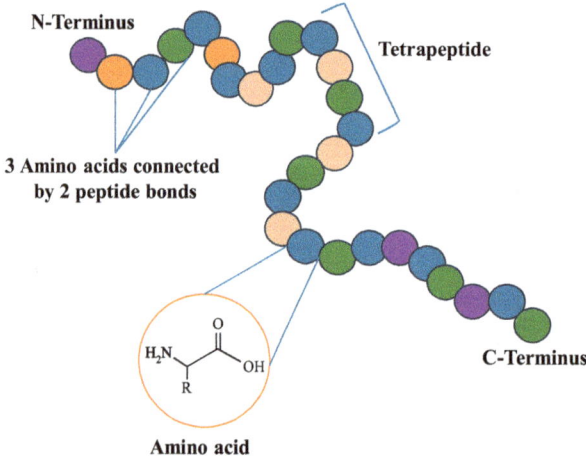

Figure 4.18: Primary protein structure.

4.7.2 Secondary Structure

Proteins may fold differently in small regions. These motifs or patterns make up secondary structures. The β-pleated sheet and the alpha helix are two common secondary structural features (Figure 4.19). These structures require intramolecular interactions to maintain their three-dimensional shape. Specifically, hydrogen bonding between the carbonyl functional groups and the amine backbone is required. Every winding turn of an alpha helix contains many amino acid residues. The polypeptide's α-helix is where the R groups stretch. The "pleats" of the β-pleated sheet are generated by hydrogen bonding between atoms on the polypeptide chain's backbone. The R groups extend above and below the pleat folds and are linked to the carbons.

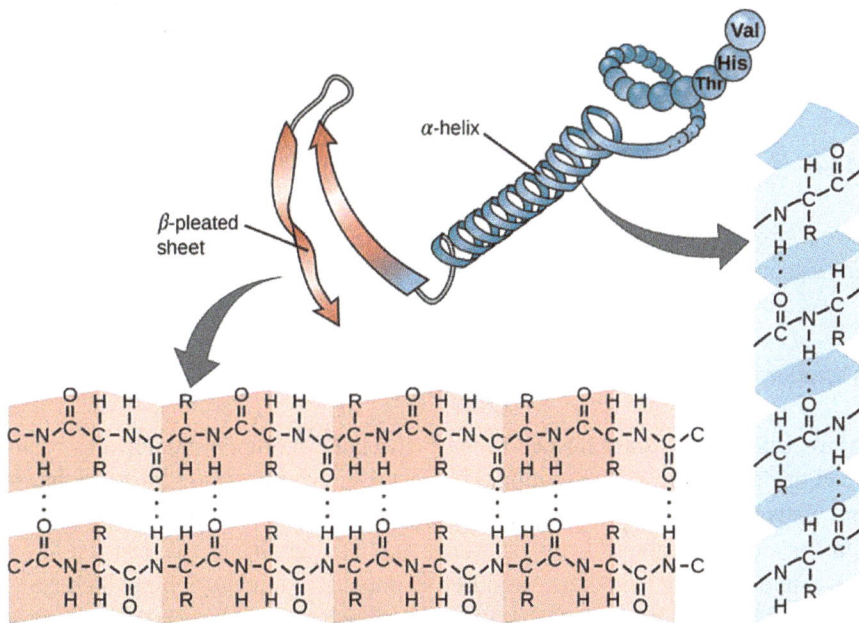

Figure 4.19: Secondary protein structure includes alpha helix and β-pleated sheet (image credit: creative commons. https://creativecommons.org/).

4.7.3 Tertiary Structure

The tertiary structure refers to the protein's overall three-dimensional shape (Figure 4.20). A wide range of chemical interactions determine proteins' tertiary structures. These include hydrophobic interactions between the hydrophobic R groups of nonpolar amino acids such as valine and leucine, ionic bonding between amino acids with opposite charges such as aspartic acid and lysine, hydrogen bonding between

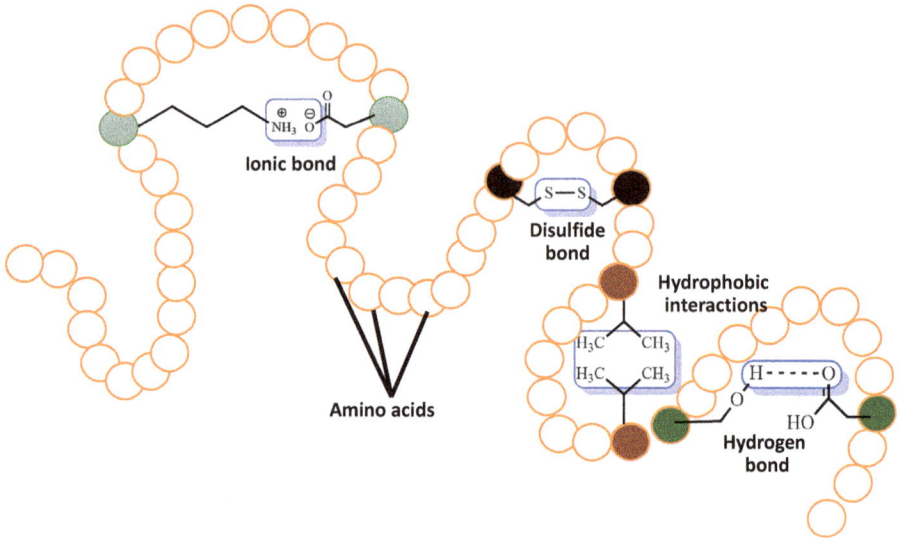

Figure 4.20: Key bonds present in the tertiary protein structure.

serine's hydroxyl group and aspartic acid's carbonyl group, and disulfide linkages between cysteine side chains.

4.7.4 Quaternary Structure

In nature, some proteins are formed from several polypeptides, also known as subunits, and the interaction of these subunits forms the quaternary structure. Weak interactions between the subunits help to stabilize the overall structure. For example, insulin has a combination of hydrogen bonds and disulfide bonds that cause it to be mostly clumped into a ball shape. Another example is the hemoglobin which composed of four protein chains, two alpha chains and two beta chains, each with a ring-like heme group containing an iron atom (Figure 4.21).

4.8 Common Proteins

4.8.1 Keratin

Keratin is one of several structural fibrous proteins known as scleroproteins. Keratin's (Figure 4.22) fundamental structure consists of α-helices and β-sheets (Figure 4.23). α-Keratin can be found in mammals' hair, skin, horns, and nails. β-keratin is found in birds' and reptiles' feathers, claws, beaks, and scales. Keratin fibers exhibit elasticity

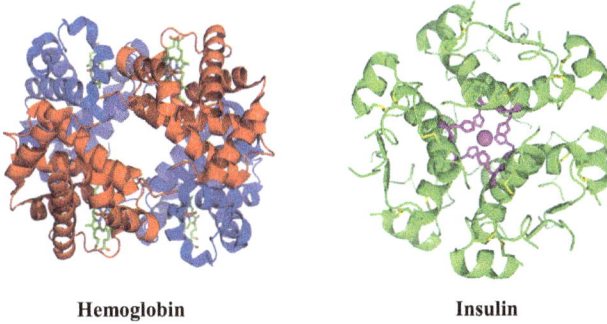

Hemoglobin   Insulin

Figure 4.21: Quaternary protein structure examples "hemoglobin, and insulin" (image credit: creative commons. https://creativecommons.org/).

due to the interplay of α-helices and β-sheets in the protein structure. Keratin is highly insoluble in both water and organic solvents. Keratin monomers bind together to produce intermediate filaments, which are tough and form strong unmineralized epidermal appendages found in reptiles, birds, amphibians, and mammals.

A keratin protein's main structure consists of amino acid chains. The amount and order of amino acids in these chains vary, as do their polarity, charge, and size. Keratins were divided into two groups based on their structure, function, and regulation: "Hard" keratins, which form ordered filaments embedded in a cysteine-rich protein

Figure 4.22: AI-generated keratin-protein structure (image credit: Easy-Peasy.AI – https://easy-peasy.ai/ai-image-generator/images/keratin-protein-structure-scientific-illustration).

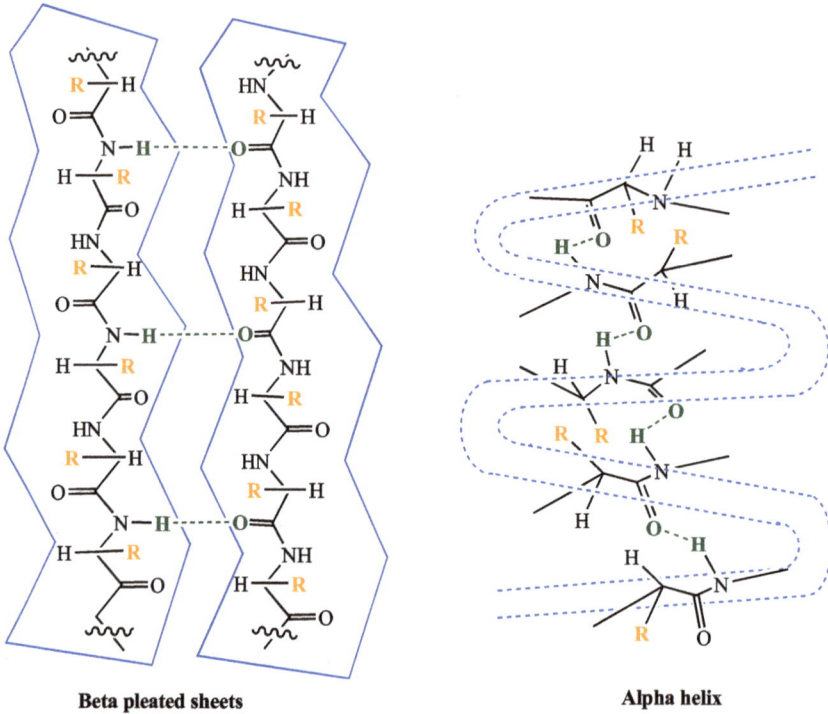

Beta pleated sheets   **Alpha helix**

Figure 4.23: Simplified secondary structures of keratin protein (β-pleated sheets and alpha helix).

matrix, presenting a compact and hard structure; and "Soft" keratins, which form loosely packed bundles of filaments and serve to grant elongation and stress release.

4.8.2 Wool

Wool is one of the first natural protein fibers. Wool is the second most extensively used natural fiber behind cotton and has been used since the beginning of human society. Wool's polymeric structure is highly complicated. Wool fiber, a cross-linked keratin protein, contains up to 20 different amino acids including glycine, alanine, valine, leucine, isoleucine, phenylalanine, proline, serine, threonine, tyrosine, aspartic, glutamic, arginine, lysine, histidine, tryptophan, cysteine, and methionine (Figure 4.24).

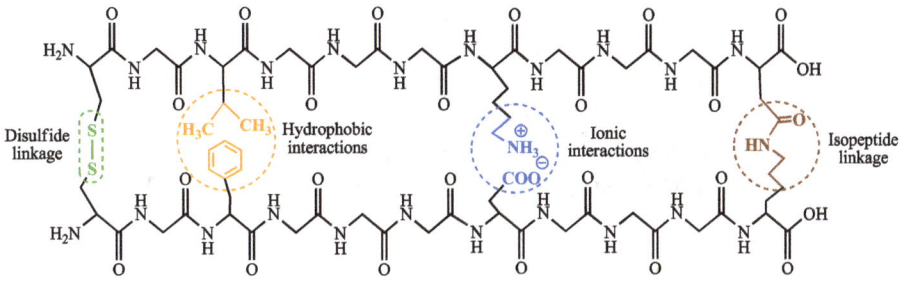

Figure 4.24: Simplified chemical structure of amino acid sequence of wool protein.

4.8.3 Silk

The structure of silk is made up of 16 different amino acids bonded together in a continuous chain. Three of these 16 amino acids, alanine, glycine, and serine, account for approximately four-fifths of the silk polymer composition (Figure 4.25). Silk is only found in beta form and contains no sulfur-containing amino acids. The raw silk strand comprises two silk filaments wrapped in a sericin protein. The raw silk strand's thickness and its uneven and irregular surface are related to the sericin coating, which gives raw silk a coarse handle. Silk is considered more plastic than elastic due to its highly crystalline polymer system, which does not allow for the same degree of polymer movement as a more amorphous system. Silk contains a very crystalline polymer structure. The increased crystallinity of silk's polymer structure allows fewer water molecules to enter. Silk is more delicate to heat due to the absence of covalent cross-links in the polymer system.

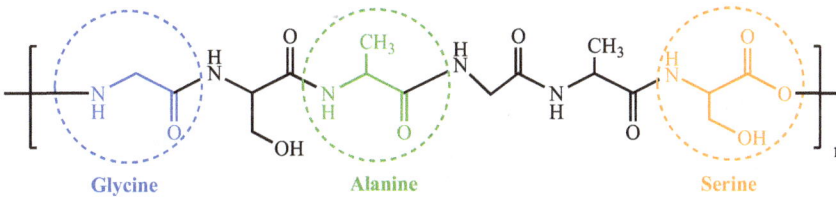

Figure 4.25: Simplified chemical structure of amino acid sequence of silk protein.

4.8.4 Collagen

Collagen is an abundant structural protein found in all mammals. In humans, collagen accounts for one-third of total protein, three-quarters of dry skin weight, and is the most common extracellular matrix component. While there are 28 identified types of collagens, the body's supply consists mostly of 5 types: Type I, Type II, Type III, Type

IV, and Type V collagen. Type I and III collagen are the most abundant in the body. There are various sources of collagen including bovine collagen, marine collagen, vegan collagen, and hydrolyzed collagen.

Collagen's amino acid sequence is commonly a repeated tripeptide unit that follows the pattern Gly-Pro-X or Gly-X-Hyp, whereby Gly is glycine, Pro is proline, Hyp is hydroxyproline, and X can be any of many amino acid residues (Figure 4.26).

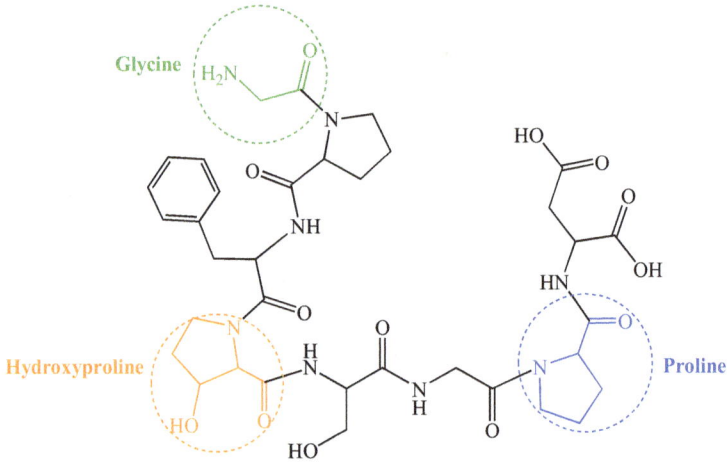

Figure 4.26: Simplified chemical structure of amino acid sequence of collagen protein.

4.8.5 Gelatin

Gelatin is an animal protein produced by thermally denaturing collagen, which is removed from animal skin and bones with a dilute acid. It could also be extracted from fish skins. Gelatin is also characterized by the triplet Gly-Pro-X or Gly-X-Hyp, whereby Gly is glycine, Pro is proline, Hyp is hydroxyproline, and X can be any amino acid residue commonly found in gelatin. Glycine makes up one-third of the chain, with proline or hydroxyproline accounting for the other third. Gelatin's polypeptide chain may also contain positively charged amino acid residues like lysine and arginine, negatively charged amino acid residues like glutamic and Asp, and hydrophobic residues like leucine, isoleucine, methionine, and valine (Figure 4.27).

4.9 Enzymes

Enzymes are exceptional biocatalysts that exist in the form of protein or RNA (ribozyme). They accelerate biochemical reactions under moderate temperature and pressure

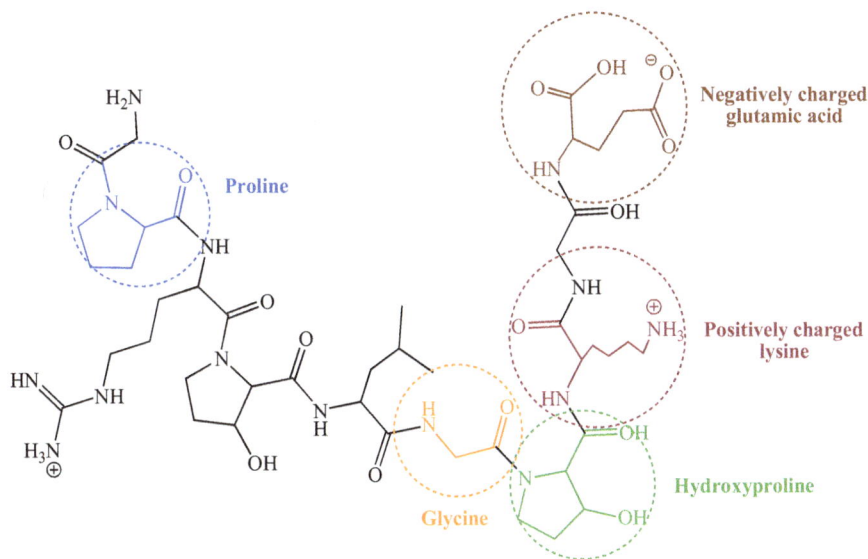

Figure 4.27: Simplified chemical structure of amino acid sequence of gelatin protein.

conditions, and practically all biochemical reactions require their presence to occur. The specificity of an enzyme is determined by the substrate's molecular geometry; consequently, the specific enzyme can catalyze one or more particular classes of processes.

The rate of any biological process is determined by a variety of parameters including temperature, pH, enzyme, and substrate concentration. Gradual temperature increase enhances the reaction rate until it reaches the optimal temperature range of 25–38 °C. Increased temperature beyond its optimum range causes a decrease in reaction rate due to enzyme denaturation, and as a result, it loses structural integrity.

4.9.1 Enzyme Structure

Enzymes are classified into two types: simple and conjugated enzymes. A simple enzyme is solely composed of proteins (amino acid chains). A conjugated enzyme is made up of two parts: a nonprotein cofactor and a protein called an apoenzyme. The apoenzyme and cofactor combine to form the holoenzyme, which is a biochemically active enzyme.

4.9.2 Enzyme Nomenclature and Classification

The Three important aspects of naming the enzyme are:
- The suffix -ase identifies the enzyme. For example, suc**rase** and lip**ase**.
- The prefix which usually the type of the reaction catalyzed by an enzyme. For example, **oxidase** enzyme catalyzes an oxidation reaction, and a **hydrolase** enzyme catalyzes a hydrolysis reaction.
- Substrate identity is also added to the type of reaction. For example, **glucose** oxidase and **pyruvate** carboxylase.

Enzymes can be categorized into six major types based on the reactions they catalyze:
- An oxidoreductase catalyzes an oxidation-reduction reaction.
- A transferase catalyzes the transfer of a functional group from one molecule to another.
- A hydrolase catalyzes a hydrolysis reaction by the addition of a water molecule.
- A lyase catalyzes the addition of a group to a double bond or the removal of a group to form a double bond.
- An isomerase catalyzes the isomerization of a substrate in a reaction.
- A ligase catalyzes the bonding together of two molecules into one with the involvement of ATP.

4.9.3 Models of Enzyme Action

4.9.3.1 Lock-and-Key Model

In the lock-and-key concept, the active site geometry is fixed, and only substrates with a complementary shape can fit into the site. According to the lock and key paradigm, an enzyme's active site has a specific conformation that precisely complements the substrate, allowing the substrate to fit into a specific place in a lock and key-like manner (Figure 4.28). The model considers the binding of a pentagon-shaped substrate to an enzyme active site with a complementary shape, which creates an enzyme-substrate complex and results in the release of a hexagon-shaped product from the enzyme. Both the enzyme and the substrate have fixed confirmations, allowing them to readily fit together. The enzyme, which can accept one or a few substrates, resembles a "lock," while the substrates represent "keys." The substrate has a complementary form to its substrate, allowing the enzyme and substrate to fit together like a "key" into a "lock." In other words, the enzyme active has a geometric shape and orientation that complements that of its substrate. The shapes of the enzyme and substrate do not affect one another. Because both enzyme and substrate already have a completely complementary structure, the substrate remains stable.

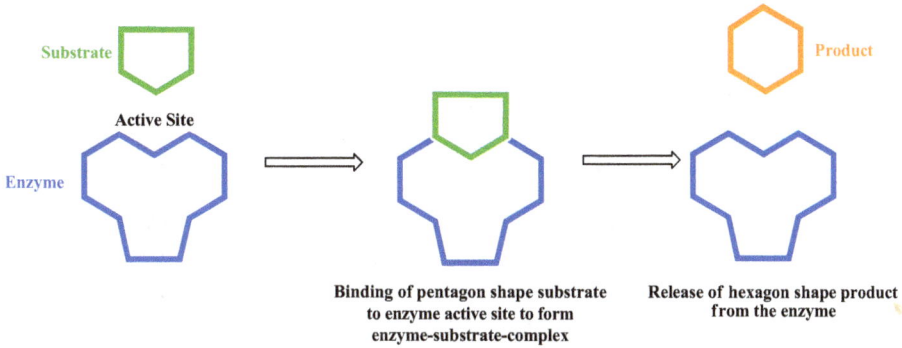

Figure 4.28: Lock-and-key model.

4.9.3.2 The Induced-Fit Model

According to the induced fit model, an enzyme's active site is not a perfect fit for its substrate. Instead, when exposed to a substrate, the active site undergoes structural changes that promote binding (Figure 4.29). According to the hypothesis of this model, the interaction between an enzyme and its substrate is analogous to putting a hand into a glove, with the moderately flexible enzyme acting as the glove and the moderately flexible substrate (the hand) fitting into it. The model proposes three crucial points: enzyme activity requires correct orientation of the catalytic group, changes in the three-dimensional connection of the active site residues may be caused by the substrate, and these changes will be triggered by substrate binding.

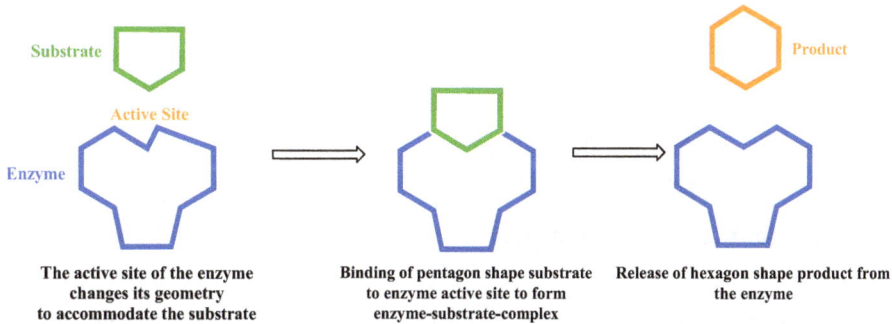

Figure 4.29: The induced-fit model.

4.9.4 Enzyme's Active Site

Seven essential characteristics define an enzyme's active site:
- The active site is merely a small portion of the enzyme. The greater component of the enzyme's structure serves as a scaffold, shaping and stabilizing the active site structure.
- The active site's complimentary form with the substrate enables a suitable fit during binding.
- The active site has catalytic residues that bind to the substrate and hold it during the chemical reaction. This binding is reversible and involves noncovalent interactions such as hydrogen bonds, van der Waals forces, and hydrophobic interactions.
- The active site avoids unwanted reactions by positioning reactants in optimal orientation and proximity to each other.
- The active site stabilizes the structure by lowering its energy during transitions.
- When interacting with a substrate, the active site can undergo considerable conformational changes.
- Active sites provide a mostly nonpolar microenvironment. Water molecules can only be found in an active site if they are reactants in the reaction.

4.10 Essential Keywords

Amino acids Organic compounds containing an amino group and a carboxylic acid group attached to carbon atom.

Collagen An abundant structural protein found in all mammals.

Electrophoresis A method depends on the movement of charged particles in an electric field used to separate and purify amino acids.

Enzymes Protein molecules that act as catalysts. Enzymes have names that provide information about their function. Most enzyme names have the suffix-ase.

Enzyme activity A measure of the rate at which an enzyme converts the substrate to products.

Enzyme's active site A small part of the enzyme in which the substrate binds.

Gelatin An animal protein produced by thermally denaturing collagen.

Induced-fit model A model where the active site can undergo small changes in geometry to accommodate the substrate.

Isoelectric point (pI or IEP) A pH at which amino acid carries no net electrical charge.

Keratin A structural fibrous protein consists of α-helices and β-sheets.

Lock-and-key model A model in which the active site of the enzyme has a fixed geometry. Only substrates with a complementary geometry can fit.

Peptide bond A chemical bond that is formed between two molecules of amino acids when the carboxyl group of one molecule reacts with the amino group of the other molecule, releasing a molecule of water.

Proteins and polypeptides A large biomolecule made of amino acid residues is linked together by peptide bonds.

Protein's primary structure A basic structure that determines protein's unique folding pattern.

Protein's secondary structure Motifs or patterns formed as a result of protein different folding in small regions.

Protein's tertiary structure A structure refers to the protein's overall three-dimensional shape.

Protein's quaternary structure A structure formed as a result of the interaction of protein's subunits.

Solid-phase peptide synthesis An automated synthetic method or technology used for the production of synthetic peptides.

Strecker synthesis A reaction that involves treating an aldehyde with ammonia and hydrogen cyanide to produce an α-aminonitrile, which is then hydrolyzed to produce the α-amino acid.

Transamination A reaction of moving an amino group from one molecule to another.

Zwitterion A molecule that has a positive charge on one atom and a negative charge on another atom.

4.11 Practice Exercises

4.11.1   Why all amino acids (except glycine) are chiral?
4.11.2   Draw the chemical structure of leucine zwitterion.
4.11.3   Draw the general chemical structure of serine and show how a peptide bond is formed between two serine molecules.

4.11.4 Draw chemical structures of alanine and tyrosine zwitterions.

4.11.5 Draw the general structure for L-amino acids.

4.11.6 Which one of the following amino acids is achiral?

    i. Lysine

    ii. Alanine

    iii. Glycine

4.11.7 Which of the following amino acids has its α-carbon as part of a five-membered ring?

    i. Proline

    ii. Tryptophan

    iii. Histidine

4.11.8 Which of the following amino acids has a nonpolar side chain?

    i. Serine

    ii. Phenylalanine

    iii. Asparagine

4.11.9 Provide the Fischer projection of L-serine.

4.11.10 How many peptide bonds are there in the peptide chain below?

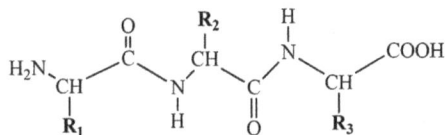

4.11.11 The isoelectric point is important in which of the following:

    i. Electrophoresis

    ii. Determination of the C-terminal amino acid

    iii. Determination of the N-terminal amino acid

4.11.12 When a disulfide linkage is formed, the compound containing this new linkage has been

    i. Oxidized

    ii. Reduced

    iii. Hydrolyzed

4.11.13 The solid-phase method of peptide synthesis was devised by

    i. Sanger

    ii. Merrifield

    iii. Strecker

4.11.14 Which functional groups in the amino acid react to form the peptide bond?

4.11.15 What coupling reagent is commonly used in solid-phase peptide synthesis?

4.11.16 Predict the function of the following enzymes:

    i. Cellulase

    ii. Sucrase

    iii. L-Amino acid oxidase

4.11.17 What type of protein facilitates or accelerates chemical reactions?
    i.   An enzyme
    ii. A hormone
    iii. A tRNA molecule

4.11.18 What are the monomers that make up proteins called?
    i.   Amino acids
    ii. Disaccharides
    iii. Nucleotides

4.11.19 Where is the linkage made that combines two amino acids?
    i.   Between the R group of one amino acid and the R group of the second.
    ii. Between the carboxyl group of one amino acid and the amino group of the other.
    iii. Between the nitrogen atoms of the amino groups in the amino acids.

4.11.20 The alpha-helix and the β-pleated sheet are part of which protein structure?
    i.   Primary structure
    ii. Secondary structure
    iii. Tertiary structure

****

Chapter 5
Nucleic Acids

5.1 Introduction

Nucleic acid is a naturally occurring chemical component that serves as the cell's primary information carrier and directs the process of protein synthesis, determining the hereditary traits of all living things. Nucleic acids are also distinguished by their capacity to be broken down to produce phosphoric acid, sugars, and a combination of organic bases (purines and pyrimidines).

Nucleic acids are classified as deoxyribonucleic acid (DNA) or ribonucleic acid (RNA). DNA is the master blueprint for life, making up the genetic material of all free-living organisms and most viruses. RNA is the genetic material of some viruses, but it is also present in all living cells and has a range of roles including protein production. Nucleotides are the monomeric building blocks of nucleic acids (Figure 5.1). Looking at this structure, it is evident that a complete understanding of nucleic acid chemistry necessitates the expertise of the other three chemistries.

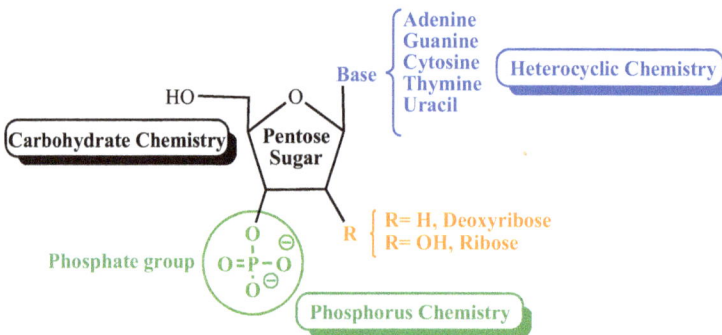

Figure 5.1: Nucleotide structure.

5.2 Structure Discovery of DNA

Swiss chemist Friedrich Miescher identified DNA for the first time in the late 1860s. Then, decades after Miescher's discovery, other scientists, most notably Phoebus Levene and Erwin Chargaff, conducted a series of studies that revealed additional details about the DNA molecule, including its primary chemical components and how they interacted with one another. Based on low-resolution X-ray crystallographic data from biophysicists Rosalind Franklin and Maurice Wilkins and Erwin Chargaff's observation that T

https://doi.org/10.1515/9783111583273-005

equals A in naturally occurring DNA and G equals C in DNA, James D. Watson and Francis H.C. Crick proposed a three-dimensional structure for DNA in 1953.

They proposed that two polynucleotide strands wrap around one another to form a double helix. Despite being identical, the two strands flow in different directions because of how the 5′–3′ phosphodiester link is oriented. The bases are located inside the helix and are connected to complementary bases on the opposite strand via hydrogen bonds. The sugar-phosphate chains run along the exterior of the helix (Figure 5.2).

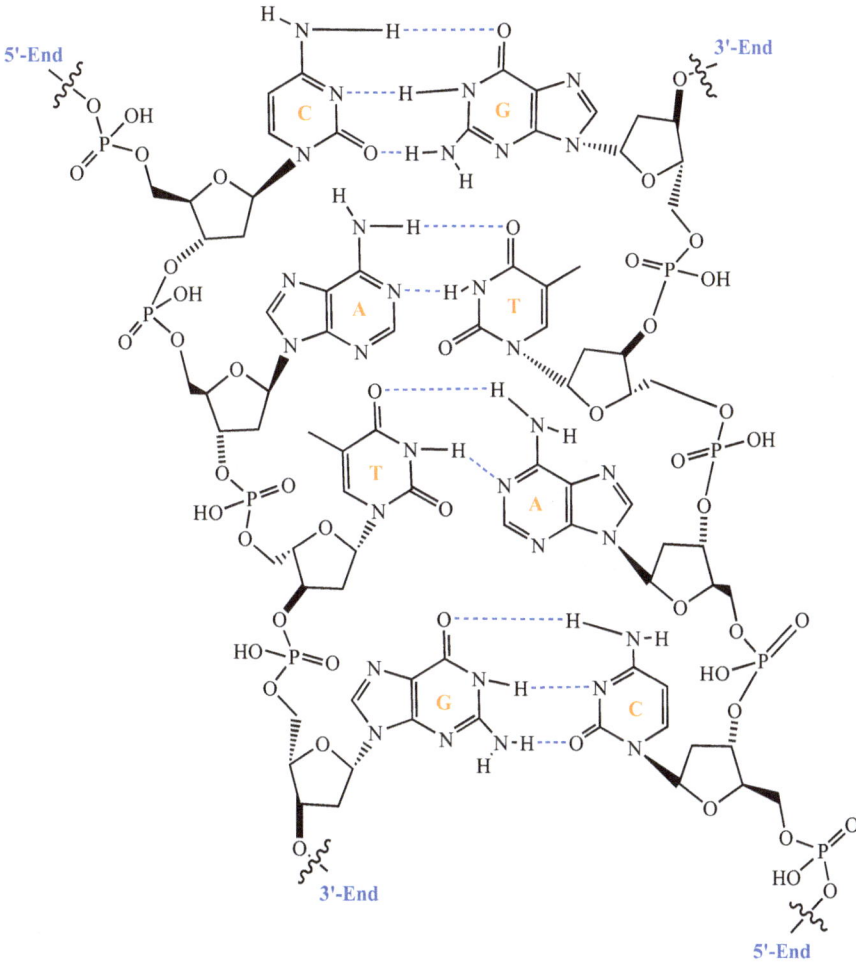

Figure 5.2: Two DNA polynucleotide strands.

5.3 Nucleic Acids Components

5.3.1 Pentose Sugars "Structure and Nomenclature"

RNA and DNA differ in the sugar unit structure in their nucleotides. In RNA the sugar unit is ribose (2-OH) and hence the R in RNA. In DNA, the sugar unit is 2-deoxyribose (2H) and hence the D in DNA (Figure 5.3).

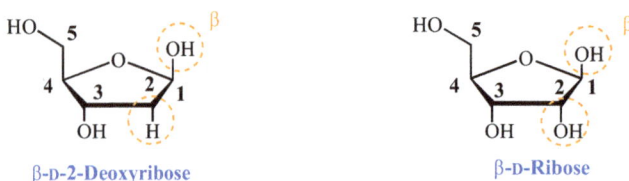

β-D-2-Deoxyribose β-D-Ribose

Figure 5.3: Pentose sugars.

5.3.2 Nucleic Acids Bases (Nitrogenous Bases or Nucleobases) "Structure and Nomenclature"

There are five primary bases (Figure 5.4A and B). Purine derivatives include adenine and guanine, while pyrimidine derivatives include thymine, cytosine, and uracil. The typical abbreviations for these five bases are A (Ade), G (Gua), T (Thy), C (Cyt), and U (Ura). Cytosine, thymine, and uracil are all numbered in the same manner. Guanine and adenine are also numbered using the same methodology.

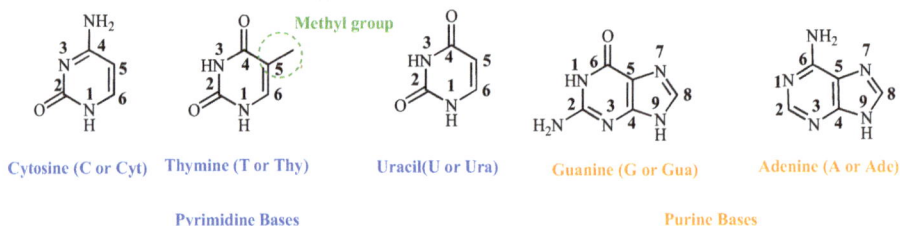

Cytosine (C or Cyt)   Thymine (T or Thy)   Uracil(U or Ura)   Guanine (G or Gua)   Adenine (A or Ade)

Pyrimidine Bases   Purine Bases

Figure 5.4A: Nucleic acid bases.

5.3.3 Nucleosides: Structure and Nomenclature

Nucleosides are produced by coupling pyrimidine or purine bases with ribose or deoxyribose sugars in the presence of appropriate reagents, as shown graphically in Figure 5.5.

Cytosine **C**

Guanine **G**

Adenine **A**

Uracil **U**

Nucleobases
of RNA

Nucleobases

Base pair

helix of
sugar-phosphates

Cytosine **C**

Guanine **G**

Adenine **A**

Thymine **T**

Nucleobases
of DNA

RNA
Ribonucleic acid

DNA
Deoxyribonucleic acid

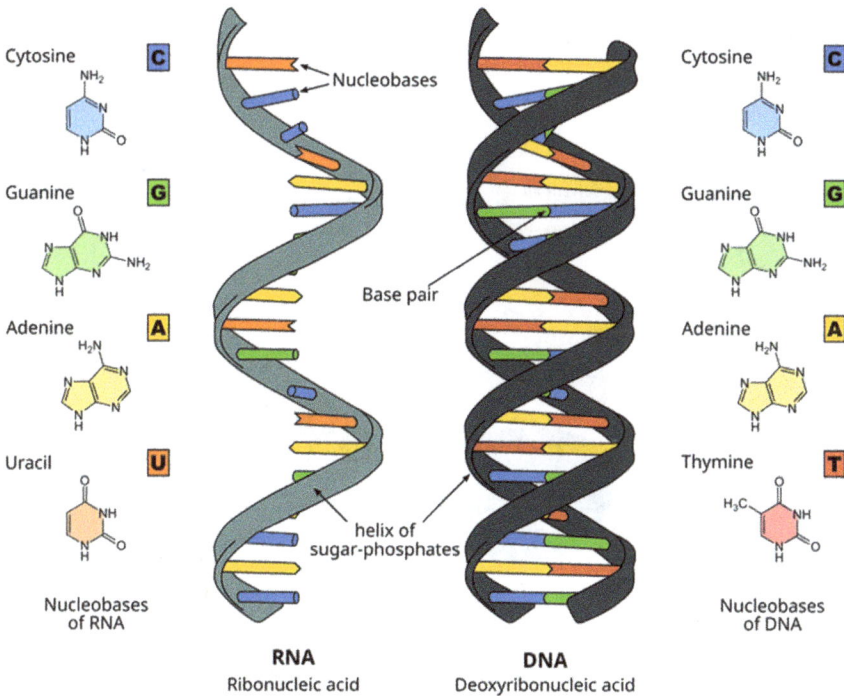

Figure 5.4B: RNA and DNA nucleobases (image credit: creative commons. https://creativecommons.org/).

D-Ribose and 2′-deoxy-D-ribose are linked to purines and pyrimidines via a β-*N*-glycosidic bond between the anomeric carbons of ribose and 2′-deoxy-D-ribose and the *N*9 or *N*1 of the purines or pyrimidines, respectively. Pentose ring atoms are identified with prime numbers. Nitrogen-containing base ring atoms are assigned unprimed numbers as shown in Figure 5.6. Selected examples of nucleosides and their components are also summarized in Table 5.1.

5.3.4 Nucleotides: Structure and Nomenclature

Each RNA nucleotide monomer comprises ribose, a phosphate group, and one of the four heterocyclic bases adenine, cytosine, guanine, or uracil. The DNA nucleotide monomer consists of deoxyribose, a phosphate group, and one of the heterocyclic bases adenine, cytosine, guanine, or thymine (Figure 5.7). The phosphate group can be added at either the 5′-OH or 3′-OH positions.

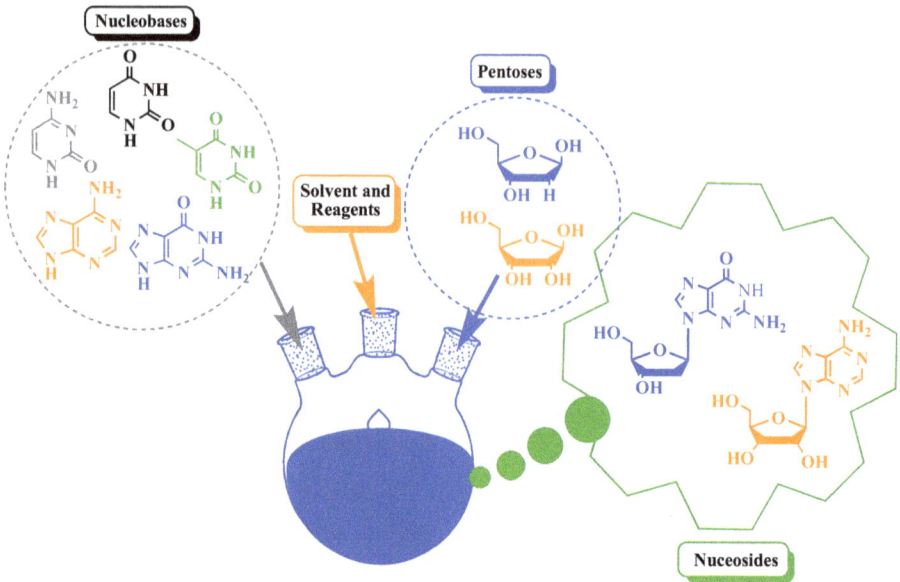

Figure 5.5: Graphical presentation of nucleoside synthesis.

5.3.5 DNA Versus RNA

There is a similarity between RNA and DNA. Both are sugar-phosphate polymers, and both have nucleobases attached to the pentose sugars but there are a few differences in structure and functions. Their comparison is shown in Figure 5.8A (graphical representation of DNA and RNA structures), Figure 5.8B (ACTG-DNA sequence and ACUG-RNA sequence), and Table 5.2.

5.3.6 Syn- Versus Anti-conformations

The nucleic acid base can exist in two distinct orientations about the *N*-glycosidic bond. These conformations are identified as **syn** and **anti**. It is the anti-conformation that predominates in nature (Figure 5.9).

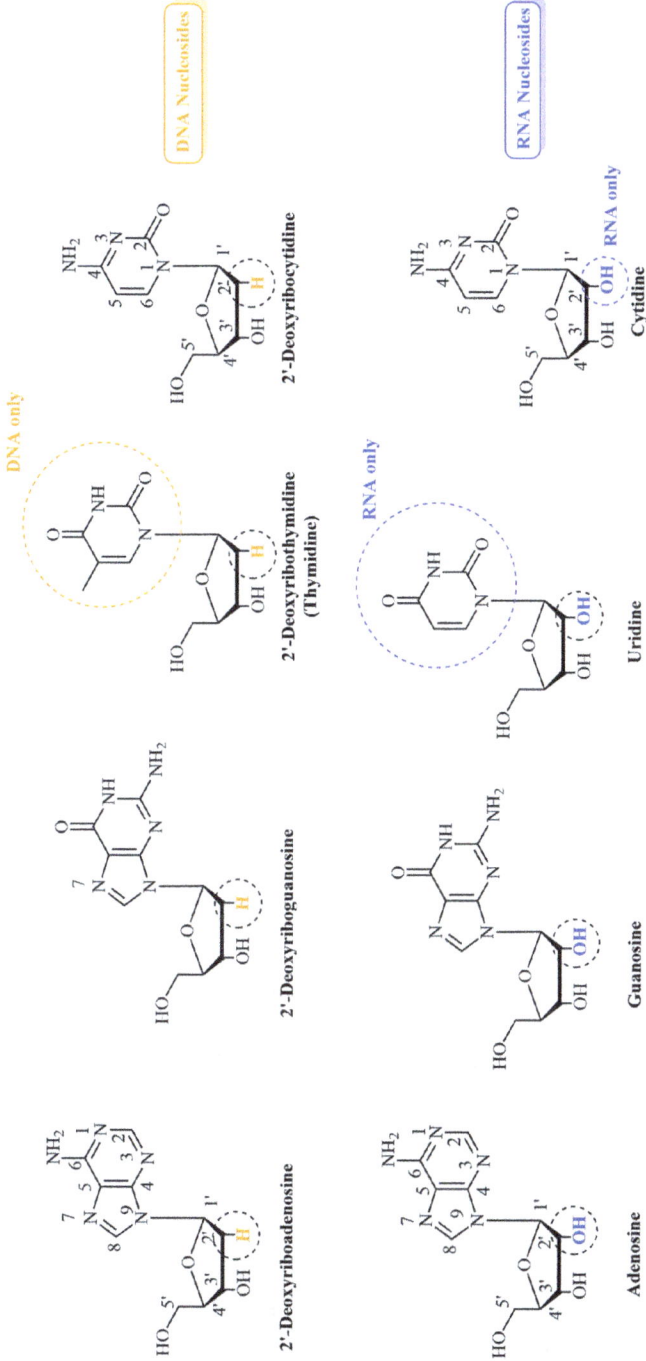

Figure 5.6: DNA and RNA nucleosides.

Table 5.1: Examples of selected nucleosides and their components.

Nucleosides	Main components
Adenosine	Adenine and ribose
Deoxyadenosine	Adenine and deoxyribose
Guanosine	Guanine and ribose
Deoxyguanosine	Guanine and deoxyribose
Cytidine	Cytosine and ribose
Deoxycytidine	Cytosine and deoxyribose
Uridine	Uracil and ribose
Deoxyuridine	Uracil and deoxyribose
Thymidine riboside	Thymine and ribose
Thymidine	Thymine and deoxyribose

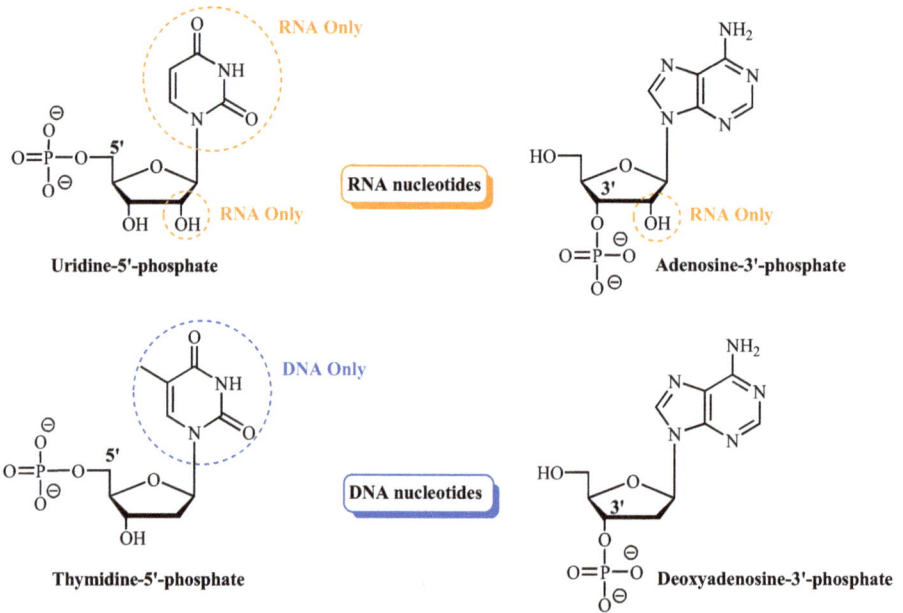

Figure 5.7: Examples of DNA and RNA nucleotides.

5.3.7 Tautomerism of the Nucleic Acid Bases

Nucleic acid bases (purines and pyrimidines) of nucleotides exist in the hydroxy pyrimidine or purine form, which can stabilize the aromatic ring or form an amide-like structure. Despite the increased stability of the hydroxy form, these bases prefer amide-like structures (Figure 5.10).

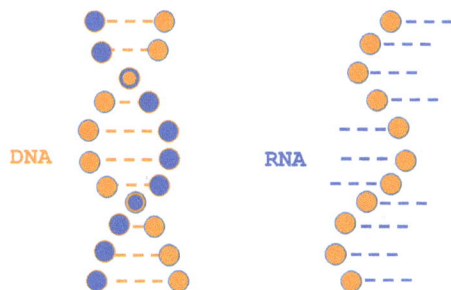

Figure 5.8A: Graphical representation of DNA and RNA structures.

5.3.8 Hydrogen Bonding in Nucleic Acids (DNA Watson-Crick Base Pairing)

Normally, the nitrogen-containing bases adenine (A) and thymine (T) pair together, as do cytosine (C) and guanine (G). The joining of these base pairs results in the double-stranded structure of DNA (Figure 5.11).

5.3.9 DNA Sugar Puckering

Sugar puckering is the process of twisting sugar molecules in DNA. This step is necessary for DNA to function properly. In the A-form, the C3'-atom is outside the plane and the sugar conformation is C3'-endo (Figure 5.12A). In the B-form, the C2'-atom is out of the plane on the same side as the nucleobase C2'-endo (Figure 5.12B). Sugar pucker shift has a wide range of consequences including duplex formation, groove widths and depths, ^intrastrand phosphate-phosphate distances, and backbone hydration.

5.4 Solid-Phase Synthesis of Oligonucleotides

The phosphoramidite monomer (Figure 5.13) acts as the foundation for the solid-phase synthesis. In this method, protective groups are added to amine, hydroxyl, and phosphate reactive sites to prevent undesirable side reactions while producing the desired final product. After the synthesis is completed, the protecting groups can be removed. The 3'-carbon is connected to the solid support, and the synthesis progresses from 3' to 5'. The solid support is made up of controlled pore glass beads with channels for connecting the protected nucleotide.

Oligonucleotide synthesis is done via a cycle of four chemical reactions that are repeated until the desired bases have been added, as shown in Figure 5.14.

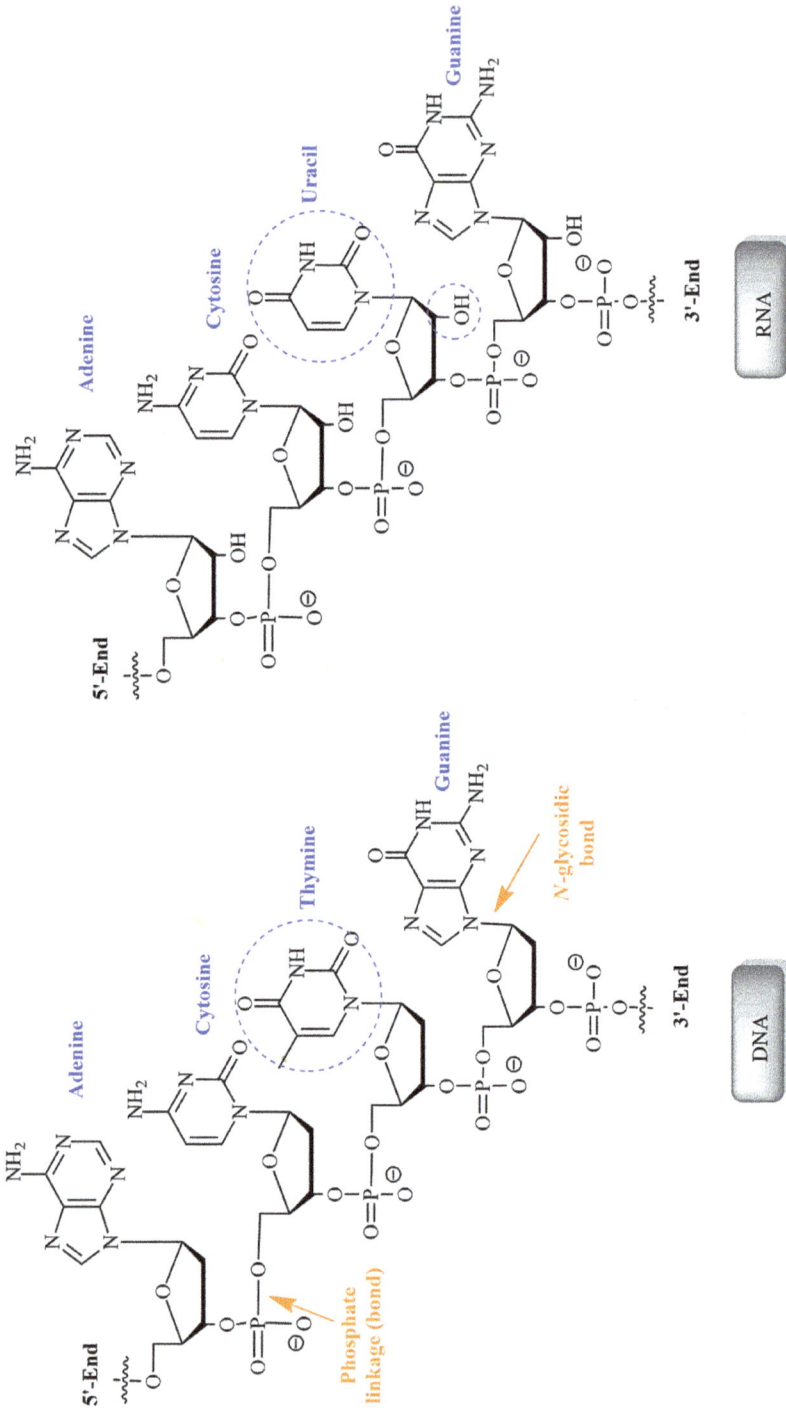

Figure 5.8B: ACTG-DNA and ACUG-RNA sequences.

Table 5.2: Comparison of nucleic acids.

Nucleic acids	Sugars	Nucleobases	Shapes	Functions
DNA	Deoxyribose	Guanine Adenine Thymine Cytosine	Double helix (double strand)	Stores genetic material.
RNA	Ribose	Guanine Adenine Uracil Cytosine	Single strand	**Messenger ribonucleic acid** (mRNA) encodes a copy of genetic information. **Transfer ribonucleic acid** (tRNA) delivers amino acids to proteins. In other words, it aids in decoding a messenger RNA (mRNA) sequence into a protein. **Ribosomal ribonucleic acid** (rRNA) is a noncoding RNA that is the primary component of ribosomes and is required by all cells. rRNA is a ribozyme that aids protein synthesis in ribosomes.

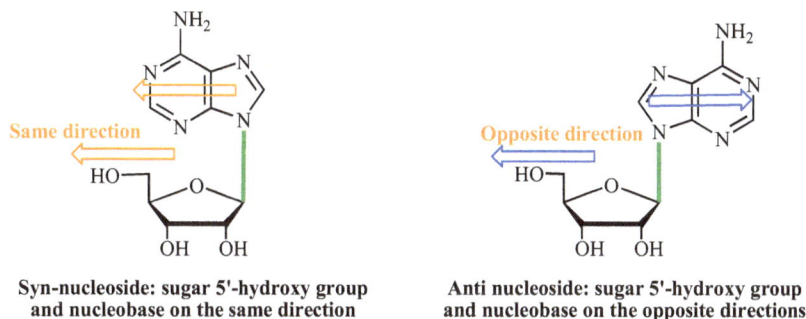

Syn-nucleoside: sugar 5'-hydroxy group and nucleobase on the same direction

Anti nucleoside: sugar 5'-hydroxy group and nucleobase on the opposite directions

Figure 5.9: Syn- versus anti-conformations in nucleic acids.

5.5 Xeno Nucleic Acids (XNAs) or Antisense Oligonucleotides (ASOs)

Nucleic acids offer fundamentally new therapeutic intervention modes (e.g., direct protein expression or targeted and programmed control of gene expression), which other biologics or small molecule drugs do not. However, several challenges have prevented nucleic acids from reaching their full potential as pharmaceuticals, including poor chemical and biological stability and limited chemical variation. To overcome these limitations, DNA and RNA analogs have been developed and tested for their ability to serve as useful biomaterials or functional molecules, encode genetic infor-

4-Amino-2-hydroxy-pyrimidine vs. cytosine   **2-Amino-6-hydroxypurine vs. guanine**

2,4-Dihydroxypyrimidine vs. uracil (R=H)
2,4-Dihydroxy-5-methyl-pyrimidine vs. -thymine (R= CH₃)

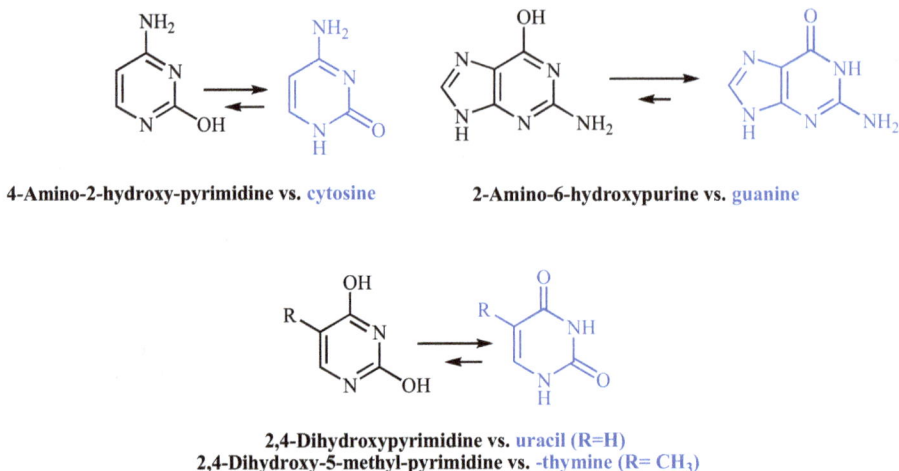

Figure 5.10: Tautomerism of the nucleic acid bases.

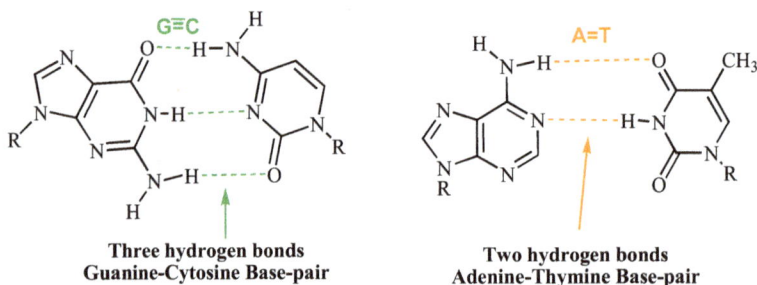

Three hydrogen bonds
Guanine-Cytosine Base-pair

Two hydrogen bonds
Adenine-Thymine Base-pair

Figure 5.11: Hydrogen bonding in nucleic acids (DNA base pairing).

mation, and drive evolution. These synthetic genetic polymers, known as xeno nucleic acids (XNAs) or antisense oligonucleotides (ASOs) have altered backbones, sugars, and nucleobases. Examples of these analogs that have sugar-backbone modifications are shown in Figure 5.15.

XNAs or ASOs are often categorized by the component of the nucleotide (sugar, backbone) carrying a modification. Shown here are GNA, glycol-glycerol nucleic acid; MNA, morpholino nucleic acid; 2′F, 2′-fluoro-nucleic acid; 2′OMe, 2′-O-methyl nucleic acid; LNA, locked nucleic acid; FANA, 2′-fluoro-arabinose nucleic acid; 2′MOE, 2′-O-methoxyethyl nucleic acid; TNA, α-L-threose nucleic acid; PS, phosphorothioate nucleic acid; phNA, alkyl phosphonate nucleic acid; PNA, peptide nucleic acid.

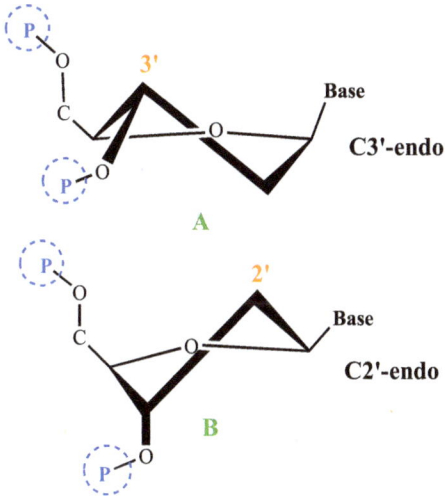

Figure 5.12: DNA sugar puckering.

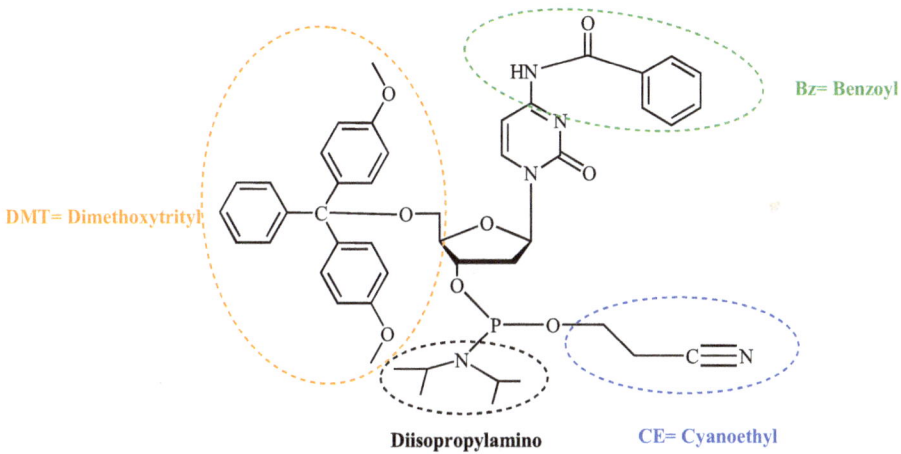

Figure 5.13: The phosphoramidite monomer.

5.6 Nucleic Acid Drugs

Nucleic acid drugs are being developed as potential treatment options. They have a high potential for treating human diseases such as cancer, viral infections, and genetic abnormalities because of their unique properties, which allow them to reach difficult targets utilizing traditional small molecule or protein/antibody-based biologics.

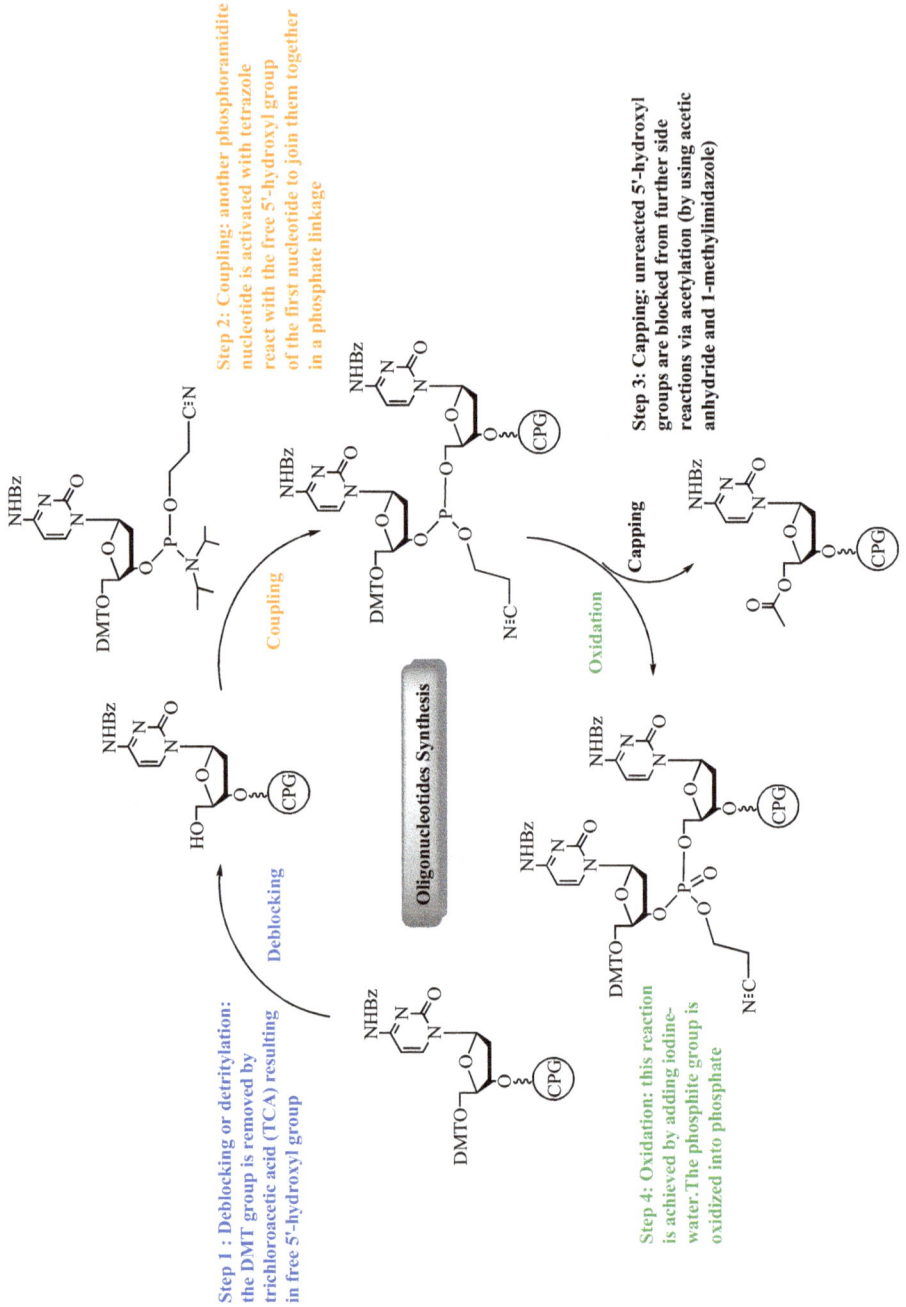

Step 1 : Deblocking or detritylation: the DMT group is removed by trichloroacetic acid (TCA) resulting in free 5'-hydroxyl group

Step 2: Coupling: another phosphoramidite nucleotide is activated with tetrazole react with the free 5'-hydroxyl group of the first nucleotide to join them together in a phosphate linkage

Step 3: Capping: unreacted 5'-hydroxyl groups are blocked from further side reactions via acetylation (by using acetic anhydride and 1-methylimidazole)

Step 4: Oxidation: this reaction is achieved by adding iodine-water. The phosphite group is oxidized into phosphate

Figure 5.14: Oligonucleotides' solid-phase synthesis.

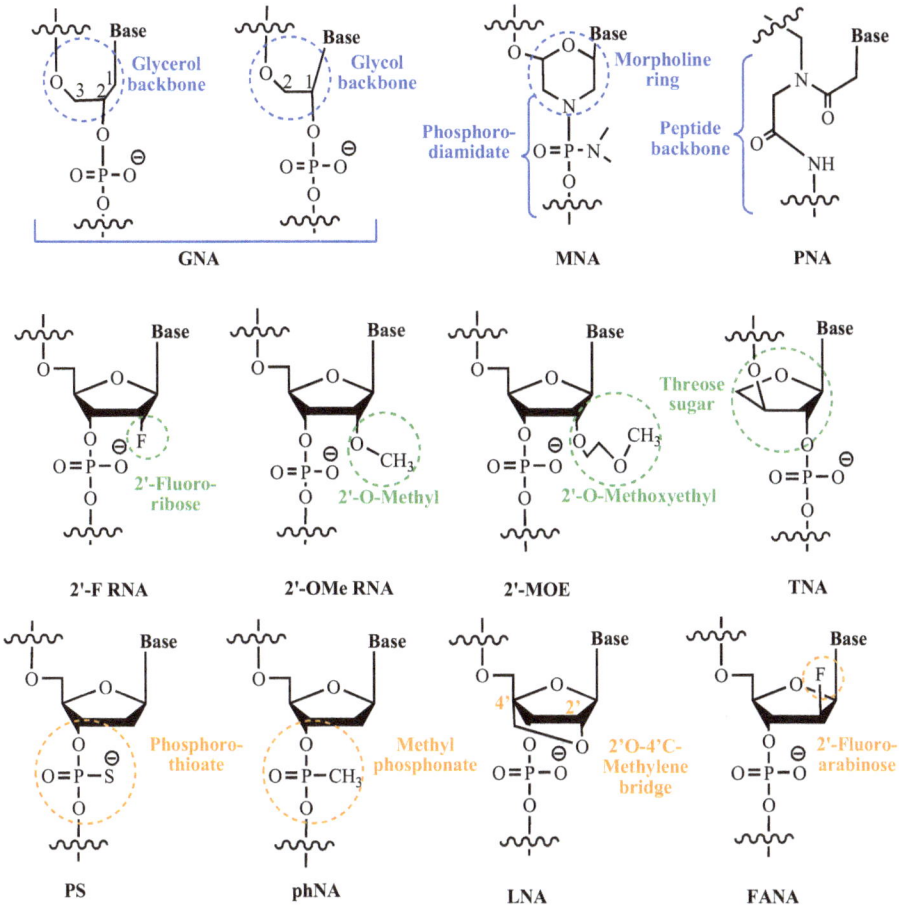

Figure 5.15: Xeno nucleic acids (XNAs) or antisense oligonucleotides (ASOs).

Nucleic acid drugs regulate cellular biological activities using nucleotide sequence information. These medications express themselves in cells or regulate genes, particularly those with complementary sequences. These methods offer a significant benefit in that nucleic acid medications can be developed regardless of the target molecule's localization or structure, enabling ways to target previously impossible molecules with small molecules or antibodies. Another advantage of nucleic acid pharmaceuticals is that once a platform is developed, a medication can be created simply by modifying the target gene's nucleotide sequence, resulting in rapid and efficient drug development. Nucleic acid drugs are categorized into five types based on how they operate on target genes (Table 5.3).

Table 5.3: Classes of nucleic acid drugs.

Therapy type	Name	Details
Inhibition	ASO	An antisense oligonucleotide (ASO) is a single-stranded, 13–30 mer DNA. An oligonucleotide with a sequence corresponding to the target gene produces a DNA/RNA double-stranded structure at the targeted location, inhibiting gene function. ASO not only creates a steric block by forming a strong double-stranded structure but also causes cleavage by RNase H.
	siRNA	siRNAs (small interfering RNA, also known as short interfering RNA or silencing RNA) are double-stranded RNAs, each 20–30 nucleotides long. When a siRNA is taken up by a cell, it forms the RNA-induced silencing complex (RISC) in the cytoplasm. Modified nucleic acids are also employed in siRNAs to inhibit nuclease degradation and prevent immunological reactions.
Splice switching	SSO	Splice-switching oligonucleotide (SSO) is a form of single-stranded DNA that controls the splicing of pre-mRNA, the precursor of mature mRNA.
Editing	EON	Editing oligonucleotides (EON) is an RNA editing technique that converts a DNA sequence at the RNA level. This method allows the rewriting of sequence information. An example is A-to-I RNA editing, which converts adenosine (A) to inosine (I) via hydrolytic deamination. Because the inosine residue on an mRNA is recognized as guanosine during translation, the editing process replaces adenosine with guanosine.
Augmentation	saRNA	Small activating RNA (saRNA) molecules increase the expression of target genes. These induce transcription from a target gene using a sequence that recognizes an upstream region of the target gene and recruits transcription factors to it on genomic DNA.
Replacement	miRNA mimic	A microRNA (miRNA) mimic is a synthetic, double-stranded RNA designed to restore a biological function that has been disrupted by a decrease in miRNA expression. This double-stranded RNA consists of a guide strand that is identical to the endogenous mature miRNA sequence and a passenger strand that is either partially or completely complementary to it.

5.7 Oligonucleotide Labeling

Many experimental methods for detecting nucleic acid function involve labeling the nucleic acid with radioisotopes or a chemical tag. Labels allow for the detection of nucleic acids, provide information about their state, and can be used as a handle to purify nucleic acids and associated components from mixtures. Radioactive molecules are frequently added to the 5′ or 3′ ends of an oligonucleotide after synthesis via enzyme-catalyzed processes. Chemical tags, on the other hand, are typically introduced during synthesis or integrated as reactive groups.

Protocols for postsynthetic attachment of chemical tags to unmodified RNA or DNA oligonucleotides include adding fluorescent dyes and biotin groups to oligonucleotides and immobilizing oligonucleotides on a solid substrate. Oligonucleotides labeled with fluorescent dyes are easily recognized in gel and plate reader experiments, whereas biotin or resin-linked oligonucleotides are beneficial for affinity purification. Fluorescent end-labeling has various advantages over radioactive labeling, including a reduction in radioactivity-related dangers and the production of a labeled molecule that does not decay, all while giving the sensitivity required for many processes.

For example, 7-diethylamino-3-[4-(iodoacetamido)phenyl]-4-methylcoumarin (DCIA) and RNA-induced silencing complex (5-IAF) can be used for 5′ labeling of DNA and RNA oligonucleotides. Aldehyde-reactive compounds such as fluorescein-5-thiosemicarbazide can be used to selectively label the 3′ end of RNA oligonucleotides. The chemical structures of these chemicals are shown in Figure 5.16.

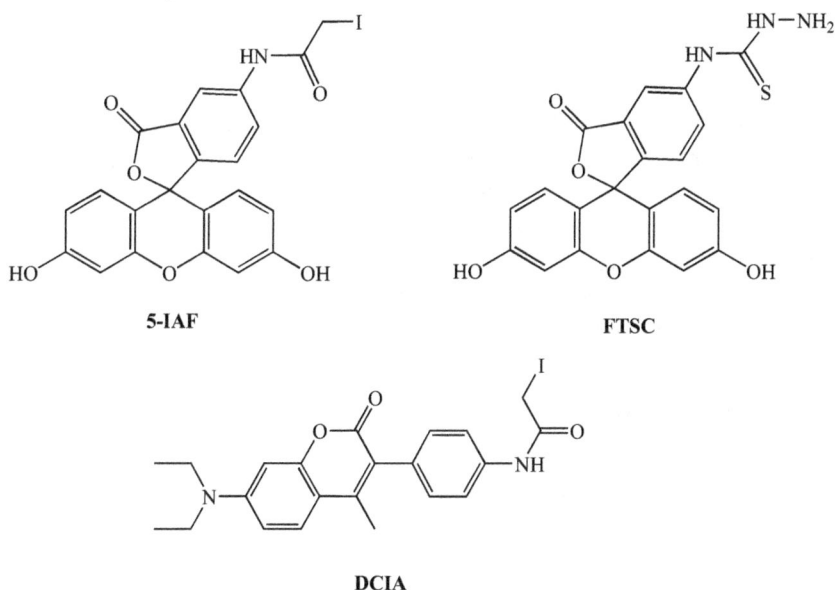

Figure 5.16: Examples of chemical tags used for labeling RNA or DNA 5′-end and RNA 3′-end.

On the other hand, the most common radioactive compounds utilized in laboratories have a short life. Phosphorus ^{32}P has a half-life of 14 days, while sulfur ^{35}S has a half-life of 68 days. Figure 5.17 illustrates general examples of these labeled compounds.

5.8 Oligonucleotide Purification and Characterization

Oligonucleotides, such as DNA and RNA, are increasingly used in diagnostic and therapeutic applications. As a result, their purity is extremely important, and purification remains difficult due to "failure sequences and final products with intact protective

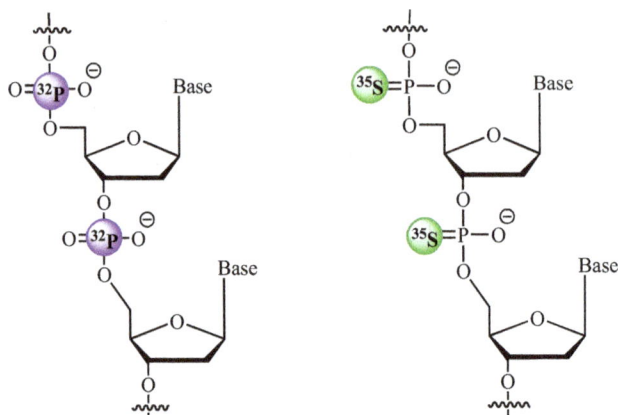

Figure 5.17: ^{32}P and ^{35}S radioactively labeled DNA.

groups." Oligonucleotide synthesis techniques such as solid-phase synthesis can introduce trace amounts of contaminants at each stage of the synthesis cycle. As the oligonucleotide length rises, the yield of the pure product drops. Modifications to oligonucleotides, such as the addition of polyethylene glycol to improve stability and bioavailability, might introduce contaminants into the production process. As the use of oligonucleotides grows, it becomes increasingly vital to develop effective methods for purifying and characterizing them. Table 5.4 summarizes selected purification and characterization approaches.

Table 5.4: Oligonucleotide purification and characterization techniques.

Technique	Brief description
	Purification
Polyacrylamide gel electrophoresis (PAGE)	A common method for separating oligonucleotides depends on size.
Ion-paired reversed-phase high-performance (IP RP HP) LC	The most prevalent method for purifying oligonucleotides. In this approach, long-chained alkyl amine is introduced at a low concentration and binds to negatively charged oligonucleotides in the LC mobile phase. The retention and elution of oligonucleotides in the LC column are influenced by parameters such as oligonucleotide charge and the length of the alkyl chain in the ion-pairing reagent such as triethylammonium acetate.

Table 5.4 (continued)

Technique	Brief description
Mass and structural characterization	
Matrix-assisted laser desorption/ionization time-of-flight (MALDI TOF) MS	This technology ionizes the oligonucleotide sample using laser light and a chemical matrix before accelerating the ions through a flight tube to a detector that tracks particles over time. The TOF is exactly proportional to the mass of the molecules.
Electrospray ionization (ESI) MS	This method uses high voltage to produce aerosol from liquid samples by ionizing target molecules into numerous charge states shown by different mass spectra, which may then be deconvoluted into parent peaks to detect oligonucleotides and potential impurities. This is a fantastic tool for analyzing oligonucleotides containing more than 50 bases.
X-ray crystallography	The most powerful method for determining the structure of oligonucleotides is X-ray crystallography. X-rays have a wavelength of roughly 1.5 Å, which is the same as interatomic bonds in molecules and can yield highly precise structures of oligonucleotides. The electron density distribution in the sample can be established by studying the scattered X-rays and reconstructing the interior molecular organization.
Nuclear magnetic resonance (NMR)	NMR is another prominent approach for characterizing oligonucleotide structures. The advantage of NMR over X-ray crystallography is that molecules are not required to be crystallized. NMR spectra give structural information based on the resonance characteristics of distinct atomic nuclei in the samples.

5.9 Central Dogma (DNA to RNA to Protein)

DNA serves as the molecular basis for heredity through the processes of replication and expression (transcription and translation). Replication yields identical DNA strands, and transcription converts DNA into messenger RNA (mRNA). Following that, mRNA is translated into amino acids by translation, which results in proteins.

5.9.1 DNA (Gene) Replication

The two strands split apart and become templates for the complementary strands during the semiconservative process of DNA replication, resulting in one parent strand and one new strand for each daughter molecule. The double strand of the ancient

DNA unzips during this process. The exposed bases are paired with free DNA nucleotides that have additional phosphates. DNA polymerase helps to link the sugars and phosphates of adjacent DNA nucleotides. Thus, as shown in Figure 5.18, two new DNA strands are created, each carrying a strand from the old DNA.

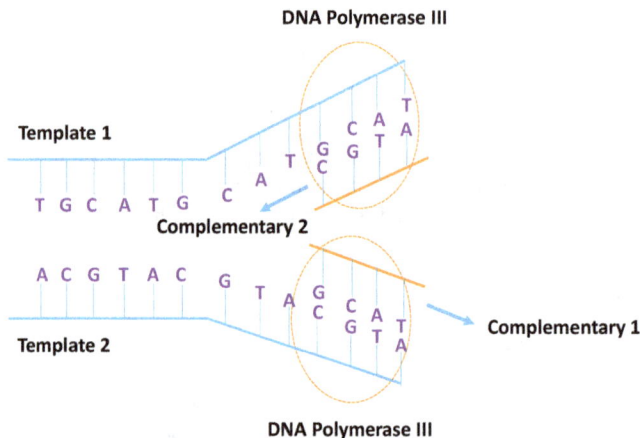

Figure 5.18: DNA (gene) replication.

5.9.2 DNA (Gene) Expression

Gene expression is primarily produced by two processes: transcription and translation. The process of transferring information from a gene's DNA to RNA, a molecule that is identical and is located in the cell nucleus, is known as transcription. The second stage of gene expression, known as translation, is when the proteins are made using the mRNAs as a template, as shown in Figure 5.19.

5.10 The Genetic Code

The genetic code refers to the instructions included in a gene that tell a cell how to manufacture a specific protein. It includes code "words" that are made up of four letters: A, C, G, and U. Adenine, cytosine, guanine, and uracil are the names of the several chemical building units known as nucleotides that each of these letters represents. As shown in the **genetic code table** (Table 5.5), the first codon nucleotide is in the left-hand column, followed by the second nucleotide in the four middle columns and the third nucleotide in the final column. Furthermore, Figure 5.20 explains the codon sequences by interpreting them from the middle of the **genetic code wheel**.

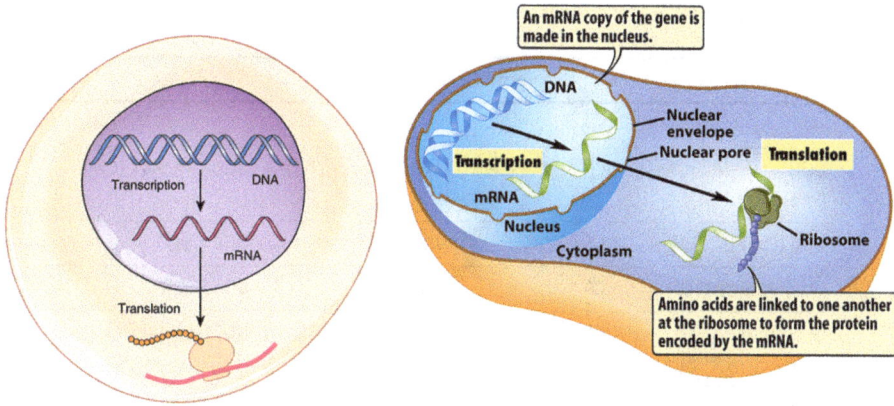

Figure 5.19: DNA (gene) expression (image credit: creative commons. https://creativecommons.org/).

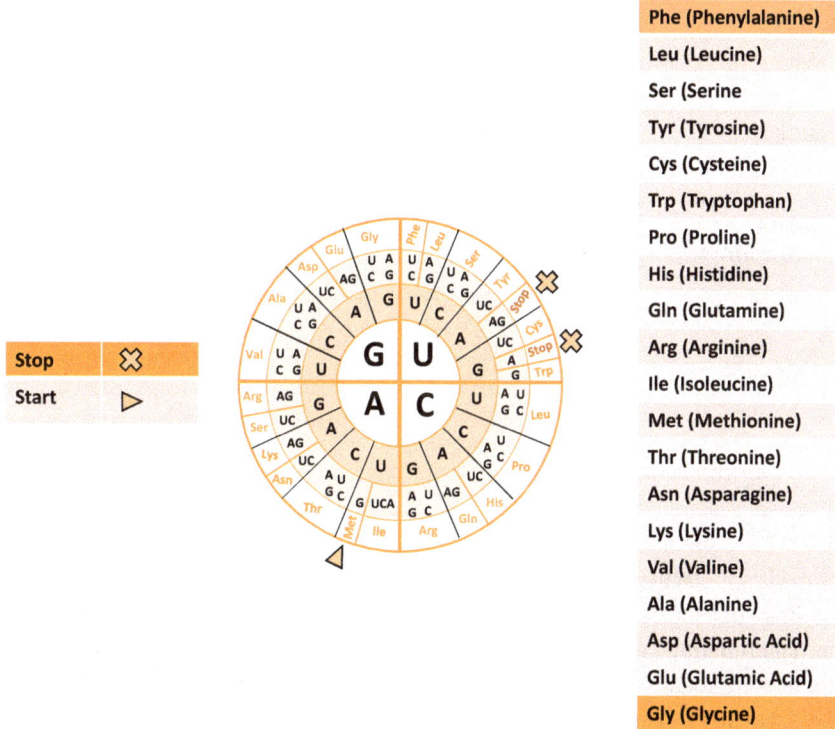

| Phe (Phenylalanine) |
| Leu (Leucine) |
| Ser (Serine |
| Tyr (Tyrosine) |
| Cys (Cysteine) |
| Trp (Tryptophan) |
| Pro (Proline) |
| His (Histidine) |
| Gln (Glutamine) |
| Arg (Arginine) |
| Ile (Isoleucine) |
| Met (Methionine) |
| Thr (Threonine) |
| Asn (Asparagine) |
| Lys (Lysine) |
| Val (Valine) |
| Ala (Alanine) |
| Asp (Aspartic Acid) |
| Glu (Glutamic Acid) |
| Gly (Glycine) |

Stop ✕

Start ▷

Figure 5.20: Genetic code wheel.

Table 5.5: Genetic code table.

First nucleotide	Second nucleotide: U	C	A	G	Third nucleotide
U	UUU Phenylalanine (Phe)	UCU Serine (Ser)	UAU Tyrosine (Tyr)	UGU Cysteine (Cys)	U
	UUC (Phe)	UCC (Ser)	UAC (Tyr)	UGC (Cys)	C
	UUA Leucine (Leu)	UCA (Ser)	UAA **STOP**	UGA **STOP**	A
	UUG (Leu)	UCG (Ser)	UAG **STOP**	UGG Tryptophan (Trp)	G
C	CUU Leucine (Leu)	CCU Proline (Pro)	CAU Histidine (His)	CGU Arginine (Arg)	U
	CUC (Leu)	CCC (Pro)	CAC (His)	CGC (Arg)	C
	CUA (Leu)	CCA (Pro)	CAA Glutamine (Gln)	CGA (Arg)	A
	CUG (Leu)	CCG (Pro)	CAG (Gln)	CGG (Arg)	G
A	AUU Isoleucine (Ile)	ACU Threonine (Thr)	AAU Asparagine (Asn)	AGU Serine (Ser)	U
	AUC (Ile)	ACC (Thr)	AAC (Asn)	AGC (Ser)	C
	AUA (Ile)	ACA (Thr)	AAA Lysine (Lys)	AGA Arginine (Arg)	A
	AUG Methionine (Met) or **START**	ACG (Thr)	AAG (Lys)	AGG (Arg)	G
G	GUU Valine (Val)	GCU Alanine (Ala)	GAU Aspartic acid (Asp)	GGU Glycine (Gly)	U
	GUC (Val)	GCC (Ala)	GAC (Asp)	GGC (Gly)	C
	GUA (Val)	GCA (Ala)	GAA Glutamic acid (Glu)	GGA (Gly)	A
	GUG (Val)	GCG (Ala)	GAG (Glu)	GGG (Gly)	G

A ribosome is a molecular machinery that reads the genetic code and translates genes into proteins. Ribosomes read three-letter words known as codons, and there are 64 different combinations of the four letters that comprise each codon. The process is made up of start codons (initiation codons include AUG and, in some cases, GUG and UUG), stop codons (termination codons include UAA, UAG, and UGA), and reading frames (the initiator AUG codon defines the reading frame of an mRNA). Translation proceeds from this start in steps of three nucleotides (one codon: a set of consecutive, nonoverlapping triplets).

For example, "AUG" encodes the amino acid methionine while also indicating the beginning of a protein. Several codons can encode the same amino acid. Because there are only 20 amino acids and 61 distinct words to encode them, there is a lot of overlapping. An amino acid can be encoded using one to six distinct codons. Only two amino acids, methionine and tryptophan, contain exactly one codon. This redundancy allows ribosomes to complete their functions correctly even when there is an error in the genetic coding.

Examples of the relationship between the DNA informational and template strand sequences along with the protein sequences for which they code are illustrated in Figure 5.21 and the three possible reading frames of the sequence AUG CAU GAC UCG UGA are also shown in Figure 5.22.

DNA informational strand   5* ATG CCA GTA GGC CAC TTG TCA 3*
DNA template strand   3* TAC GGT CAT CCG GTG AAC AGT 5*
mRNA   5* AUG CCA GUA GGC CAC UUG UCA 3*
Protein (amino acid sequence)   Met Pro Val Gly His Leu Ser

DNA informational strand   5* ATG CCA GTA GGC CAC TTG TCA 3*
DNA template strand   3* TAC GGT CAT CCG GTG AAC AGT 5*
mRNA   5* AUG CCA GUA GGC CAC UUG UCA 3*
Protein (amino acid sequence)   Met Pro Val Gly His Leu Ser

DNA informational strand   5* AAC GTT CAA ACT GTC 3*
DNA template strand   3* TTG CAA GTT TGA CAG 5*
mRNA   5* AAC GUU CAA ACU GUC 3*
Protein(amino acid sequence)   Asn Val Gln Thr Val

Figure 5.21: Relationship between the DNA informational-template sequences and protein sequence.

The sequence AUG CAU GAC UCG UGA has three different reading frames, as shown below:
              Start   Stop

Reading frame 1:
                    AUG CAU GAC UCG UGA
                    Met His   Asp Ser Stop

Reading frame 2:
                    A UGC AUG ACU CGU GA
                       Cys Met Thr Arg

Reading frame 3:
                    AU GCA UGA CUC GUG A
                       Ala Stop Leu Val

Figure 5.22: Reading frames for AUG CAU GAC UCG UGA sequence.

5.11 Polymerase Chain Reaction (PCR)

The polymerase chain reaction (PCR) is typically utilized when a greater quantity is required from a limited amount of DNA. The primary function of PCR is to generate numerous copies of a particular DNA sequence. For example, from a single picogram of DNA, several micrograms can be obtained in a matter of hours.

The PCR employs Taq DNA polymerase, which can generate the whole complementary strand from a single strand of DNA with a short primer segment of complementary chain at one end. Figure 5.23 depicts the total procedure as a series of three steps. During the process, the tiny amount of DNA to be amplified is heated in the presence of Taq polymerase, Mg^{2+} ion, the four deoxynucleotide triphosphate monomers, and a large excess of two short oligonucleotide primers of about 20 bases each. Each primer is complementary to the sequence at the end of one of the target DNA segments.

5.12 DNA Fingerprinting

DNA fingerprinting is a method of isolating and detecting variable components within the base-pair sequence of DNA. A British scientist, Alec Jeffreys, devised the approach in 1984 after noticing that certain highly variable DNA sequences (known as minisatellites) that do not contribute to gene function are repeated inside genes. Jeffreys noticed that each individual possesses a distinct set of minisatellites (10 to more than 100 base pairs in length). Except for identical twins, everyone experiences somewhat varied patterns. To create a DNA fingerprint gather a sample of DNA-containing cells such as skin, hair, or blood cells. The DNA taken from the cells is purified. Jeffreys' first approach, based on restriction fragment length polymorphism technology, involved cutting the DNA at certain locations along the strand with restriction enzymes. The enzymes produced fragments of varied lengths, which were then separated by placing them on a gel and exposing them to an electric current (electrophoresis). The shorter the fragment, the faster it migrated to the anode.

The sorted double-stranded DNA fragments were next blotted, whereby they were split into single strands and transferred to a nylon sheet (or nitrocellulose sheet). The fragments were subjected to autoradiography, which included exposure to radioactive synthetic DNA fragments called DNA probes that are attached to tiny satellites. The fragments were then subjected to a section of X-ray film, which resulted in the creation of a dark stain anywhere a radioactive probe was connected. The resulting pattern of markings might then be analyzed (Figure 5.24).

For criminal cases, the most precise 13 core STR loci (microsatellites or short tandem repeats) "STRs" feature shorter repetition units (usually 2–4 base pairs in length) and are used to identify each tested individual. Based on the 13 loci, a Combined DNA Index System is implemented to act as a registry for any convicted offenders. When a DNA sample is retrieved from a crime scene, it is cleaved with restriction endonu-

STEP 1
Denaturing: In this step, the temperature is raised to 95 °C so that the double stranded DNA denatures into two single strands.

STEP 2
Annealing: When the DNA is denatured into two single strands the temperature is then lowered to 50 °C to force the DNA primers to attach to the template DNA by hydrogen bonding to their complementary sequence.

STEP 3
Extending: When the temperature is raised to 72 °C, a new strand of DNA is then made by the Taq polymerase and when replication of each strand is complete, two copies of the original DNA are produced.

Step 1

Step 2

Step 3

Denature at 95 °C

Anneal Primers
at 50°C

Extend by
Taq polymerase,
dNTPs, and Mg2+

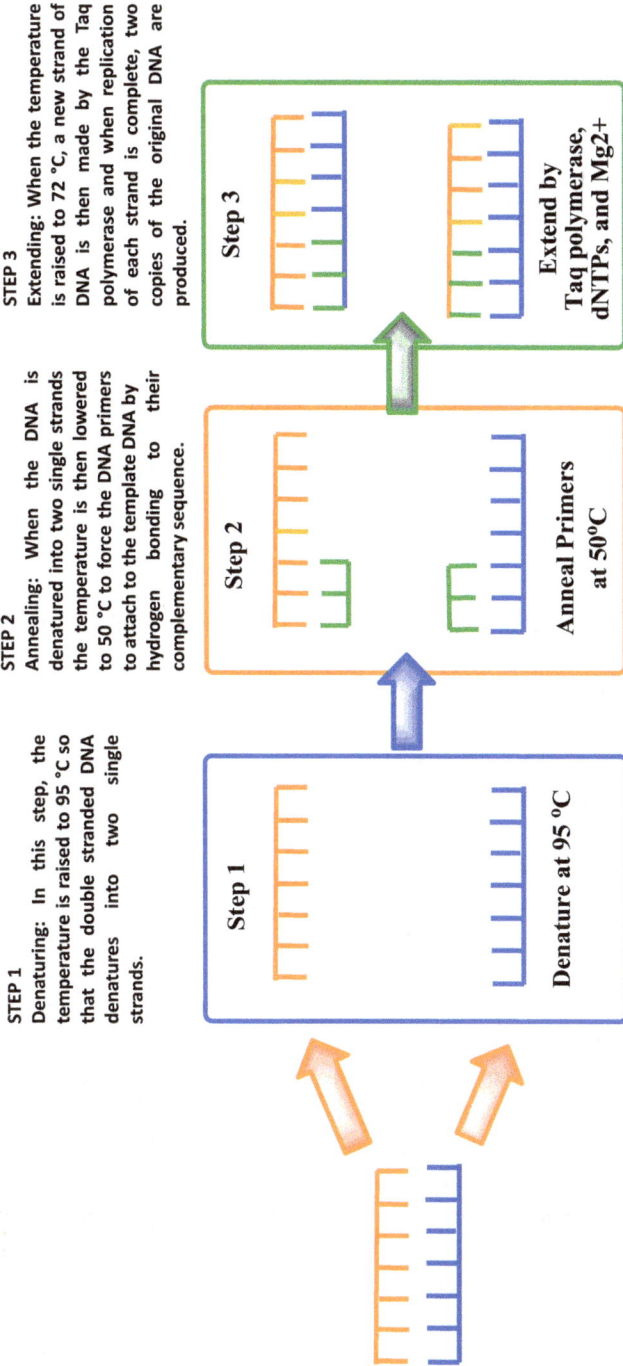

Repeating the denature–anneal–extend cycle, a second time yields four DNA copies, repeating a third time yields eight copies, and so on, in an exponential series.

Figure 5.23: PCR overall process.

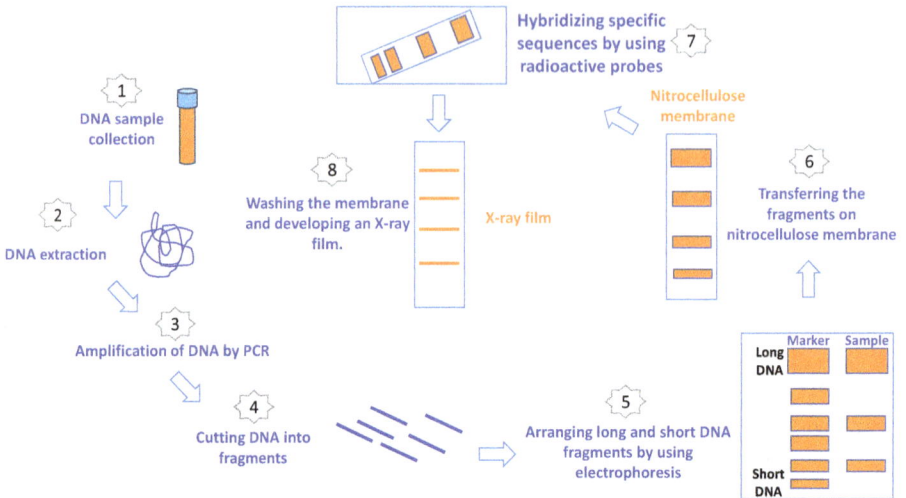

Figure 5.24: DNA fingerprinting method.

cleases to extract fragments containing the STR loci, which are then amplified using PCR and their sequences established. If the profile sequences of a known individual and DNA acquired at a crime scene match, the probability of the DNA being from the same individual is around 82 billion to one.

In addition to the above-mentioned criminal investigations, DNA fingerprinting is also used for a variety of purposes including forensic analysis and paternity testing (Figure 5.25). In these circumstances, the goal is to "match" two DNA fingerprints such as those of a known and an unknown individual.

5.13 DNA Secret Code

Every cell in our body carries DNA. DNA is a set of instructions that teaches our cells how to produce proteins. These instructions are written in a language we didn't comprehend until lately. A strand of DNA resembles a ladder. The rungs of this ladder are formed of bases. Each rung consists of two bases that are joined together in the center. The four bases used in DNA are cytosine, guanine, adenine, and thymine, which are paired in a precise order: adenine with thymine and guanine with cytosine. The arrangement of the DNA bases tells cells how to arrange amino acids. It requires three DNA bases to link with one amino acid. A codon is a three-base sequence that specifies the type of amino acid to be used. These codes can be converted into a message using the DNA code language, also known as DNA secret code (Figure 5.26). Table 5.6 shows examples of these messages.

Genetic fingerprints are inherited: 50% from mother and 50% from father, hence a person could be easily identified when fingerprints are compared between parents and siblings

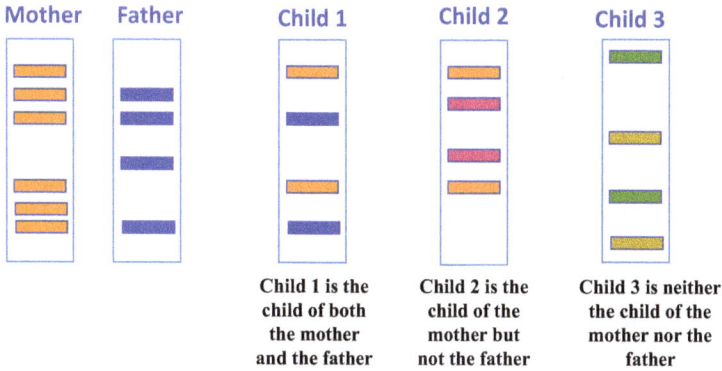

Child 1 is the child of both the mother and the father

Child 2 is the child of the mother but not the father

Child 3 is neither the child of the mother nor the father

Figure 5.25: DNA fingerprinting paternity testing.

Codon	English	Codon	English	Codon	English	Codon	English
AAA	a	CAA	q	GAA	G	TAA	W
AAC	b	CAC	r	GAC	H	TAC	X
AAG	c	CAG	s	GAG	I	TAG	Y
AAT	d	CAT	t	GAT	J	TAT	Z
ACA	e	CCA	u	GCA	K	TCA	1
ACC	f	CCC	v	GCC	L	TCC	2
ACG	g	CCG	w	GCG	M	TCG	3
ACT	h	CCT	x	GCT	N	TCT	4
AGA	i	CGA	y	GGA	O	TGA	5
AGC	j	CGC	z	GGC	P	TGC	6
AGG	k	CGG	A	GGG	Q	TGG	7
AGT	l	CGT	B	GGT	R	TGT	8
ATA	m	CTA	C	GTA	S	TTA	9
ATC	n	CTC	D	GTC	T	TTC	0
ATG	o	CTG	E	GTG	U	TTG	SPACE
ATT	p	CTT	F	GTT	V	TTT	PERIOD (.)

DNA Secret Code

Figure 5.26: DNA secret code.

Table 5.6: Examples of DNA secret code messages.

Message	Translation in English
CTC ACA AAG ATG AAT AGA ATC ACG TTG CTC GCT CGG TTG AGA CAG TTG AAAATAAAACGCAGAATCACG	Decoding DNA is amazing
CTCACATTGCACCCACGACATACACACTTGAGACAGTTGAGACAGTTGAAA ATAAAACGCAGAATCACG	De Gruyter is an excellent publisher
GTCACTAGACAGTTGAGACAGTTGGGCCACATGACCACACAGCAGATGCAC TTGCTGAGTCGCAAAACGACTACAAGAAATTTGCAGACACCCACAATCCAT ACTTTGCGTATGATGAGGTTGCCGAGACATACTTTGCACCCACGACATACA	This is Professor Elzagheid's seventh book with De Gruyter

5.14 Essential Keywords

Antisense oligonucleotide (ASO) A single-stranded, 13–30 mer DNA with a sequence corresponding to the target gene produces a DNA/RNA double-stranded structure at the targeted location, inhibiting gene function.

DNA Deoxyribonucleic acid consists of two polynucleotide strands twisted together in a double helix.

DNA fingerprinting A technique used by crime laboratories to link suspects to biological evidence blood, hair follicles, or skin found at a crime scene.

DNA secret code A language based on a codon "a three-base sequence" used for writing messages with secret code.

Editing oligonucleotides (EON) An RNA editing technique that converts a DNA sequence at the RNA level.

Genetic code Refers to the instructions included in a gene that tell a cell how to manufacture a specific protein.

Glycerol nucleic acid A chemical like DNA or RNA but differing in the composition of its backbone. GNA's backbone is composed of repeating glycerol (three carbon atoms) units linked by phosphodiester bonds.

Glycol nucleic acid A chemical like DNA or RNA but differing in the composition of its backbone. GNA's backbone is composed of repeating glycol (two carbon atoms) units linked by phosphodiester bonds.

Messenger ribonucleic acid mRNA encodes a copy of genetic information.

MicroRNA (miRNA) mimic A synthetic, double-stranded RNA designed to restore a biological function that has been disrupted by a decrease in miRNA expression.

Modified nucleic acids A distinguished nucleic acid from naturally occurring DNA or RNA by changes to the backbone of the molecule.

Morpholino nucleic acid Modified nucleic acids that have standard nucleic acid bases bound to morpholine rings instead of deoxyribose rings and linked through phosphorodiamidate groups instead of phosphates.

Nuclear magnetic resonance (NMR) A prominent approach for characterizing oligonucleotide structures.

Nucleosides Purine and pyrimidine bases are linked to D-ribose or 2′-deoxy-D-ribose.

Nucleotides The phosphate esters of nucleosides.

Oligonucleotide solid-phase synthesis Synthesis of a sequence of nucleotides by automated solid-phase techniques, where the nucleotide chain is built up by adding a protected a phosphoramidite monomer to a protected nucleotide linked to a solid phase in the presence of a coupling agent.

Polyacrylamide gel electrophoresis (PAGE) A common method for separating oligonucleotides depends on size.

Peptide nucleic acid A chemical like DNA or RNA but differing in the composition of its backbone.

Polymerase chain reaction (PCR) A technique used when a larger amount is required from available tiny amount of DNA. PCR produces multiple copies of a given DNA sequence.

RNA Ribonucleic acid consists of one polynucleotide strand.

Ribosomal ribonucleic acid rRNA molecule is a noncoding RNA that is the primary component of ribosomes and is required by all cells. rRNA is a ribozyme that aids protein synthesis in ribosomes.

Small activating RNA (saRNA) Molecules that increase the expression of target genes.

Small interfering RNA (siRNAs) Also known as short interfering RNA or silencing RNA are double-stranded RNAs with 20–30 nucleotides long. Their role is to inhibit nuclease degradation and prevent immunological reactions.

Splice-switching oligonucleotide (SSO) A form of single-stranded DNA that controls the splicing of pre-mRNA, the precursor of mature mRNA.

Syn- versus anti-conformations The orientations of nucleic acid bases around the *N*-glycosidic bond.

4'-Thionucleic acid A chemical like DNA or RNA but differing in the composition of its sugar structure. 4'-Thionucleic acid backbone is composed of repeating 4'-thio-2'-deoxyribose units linked by phosphodiester bonds.

Transfer ribonucleic acid tRNA molecule that delivers amino acids to proteins. In other words, it aids in decoding an mRNA sequence into a protein.

Threose nucleic acid A chemical like DNA or RNA but differing in the composition of its backbone. TNA's backbone is composed of repeating threose units linked by phosphodiester bonds.

X-ray crystallography The most powerful method for determining the structure of oligonucleotides.

5.15 Practice Exercises

5.15.1 Name the protecting groups in the following phosphoramidite monomer:

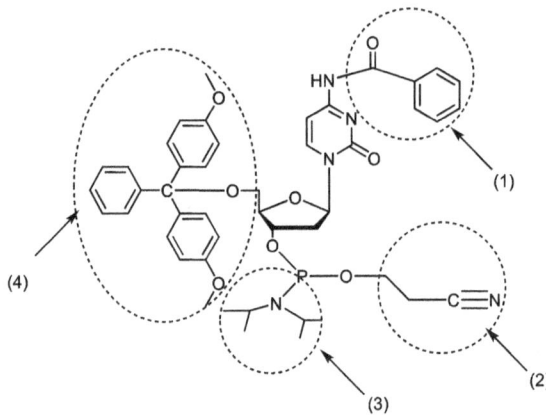

5.15.2 Write down the full names of the following artificial nucleic acids:
  i.   PNA
  ii. TNA
  iii. MNA

5.15.3 What DNA base sequence is complementary to the following DNA base sequence C-A-C-G-C-T-A-T-G-A-T-A-T-C-G-C-C-G?

5.15.4 List the three main components of nucleotides.

5.15.5 How many complementary hydrogen bonds occur when guanine and cytosine form a base pair?

5.15.6 Identify the conformation of the following nucleosides:

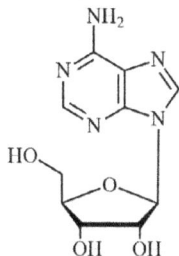

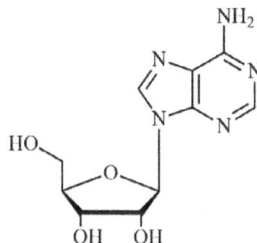

5.15.7 What are the first two steps in the oligonucleotide synthesis?

5.15.8 Assign numbers to the atoms in the structures of the following nucleic acid bases:

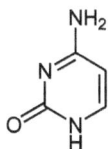

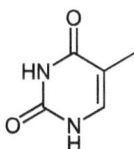

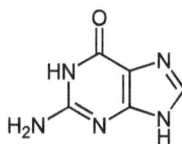

5.15.9 Give two differences between DNA and RNA components.

5.15.10 Write down the full name of the nucleic acid base G.

DNA	RNA

5.15.11 Which of the following nitrogens of adenine connects to ribose or deoxyribose to form a nucleoside?

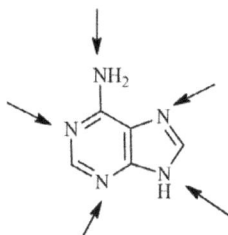

5.15.12 Which of the following nitrogens of thymine connects to deoxyribose to form a nucleoside?

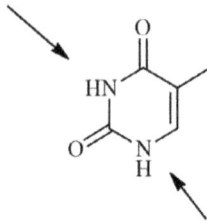

5.15.13 What are the four common ribonucleosides?

5.15.14 Name two pyrimidine bases that can exist in deoxyribonucleotides.

5.15.15 Show the hydrogen bonding that occurs when guanine and cytosine form a base pair.

5.15.16 Show the hydrogen bonding that occurs when adenine and thymine form a base pair.

5.15.17 Identify the following bases and tell whether each is found in DNA, RNA, or both:

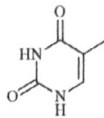

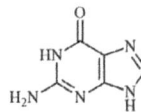

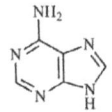

5.15.18 What amino acid sequence is coded for by the mRNA base sequence CAG-AUG-CCU-UGG-CCC-UUA?

5.15.19 Translate the following DNA secret code message into English: CTAAAAAT CAAAAATAAA.

5.15.20 Give two examples of the chemicals that can be used for 5′ labeling of DNA and RNA oligonucleotides.

****

Solutions to Practice Exercises

Chapter 1

1.8.1 A covalent bond is an interatomic connection that occurs when two atoms
       share an electron pair. The binding occurs because their nuclei are attracted to
       the same electrons. Covalent bond examples include hydrogen (H_2), oxygen
       (O_2), nitrogen (N_2), water (H_2O), and methane (CH_4).

1.8.2 1 = Monosaccharides
       2 = Proteins
       3 = Nucleotides

1.8.3 Because it is a simple and time-saving formula.

1.8.4 Acceptor Donor

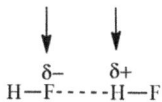

$$\text{H—F-----H—F}$$
$$\delta^- \qquad \delta^+$$

1.8.5 Water (H_2O)

1.8.6 1 = Amino group
       2 = Hydroxyl group
       3 = Sulfhydryl group
       4 = Carboxyl group
       5 = Phosphoryl group

1.8.7 Hydrogen bond donor   Hydrogen bond acceptor

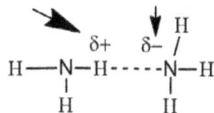

$$\text{H—N–H-----N—H}$$

https://doi.org/10.1515/9783111583273-006

1.8.8

O-Glycosidic Bond

Peptide Bond

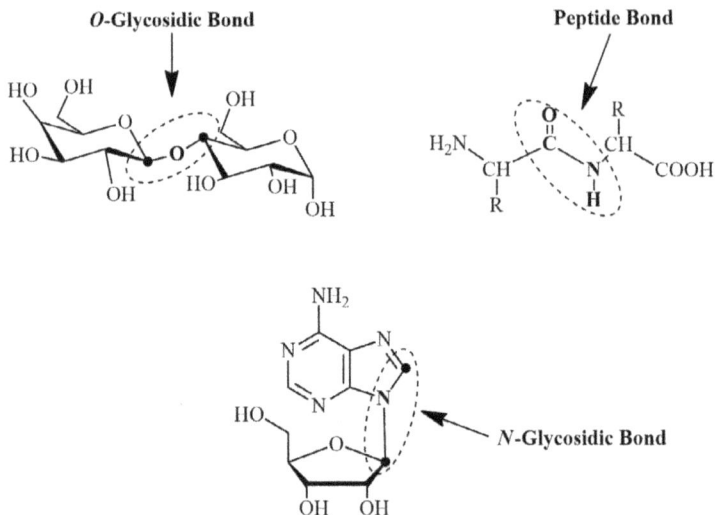

N-Glycosidic Bond

1.8.9

    i.   Five carbons

    ii. Four carbons

    iii. Three carbons

1.8.10 Hydrogen bonding is a dipole-dipole attraction between molecules that do not form a covalent connection with a hydrogen atom. It is caused by the attractive interaction between a hydrogen atom covalently bound to a very electronegative atom, such as an N, O, or F atom, and another strongly electronegative atom. Water and ammonia can form hydrogen bonds with each other.

1.8.11

    i. $3.16 - 0.93 = \textbf{2.23}$

    ii. $3.16 - 3.16 = \textbf{0}$

    iii. $2.58 - 2.20 = \textbf{0.38}$

1.8.12 Carboxylate group

            Phosphate or phosphoryl group

            Hydroxyl group

1.8.13 Deoxyribonucleic acid (DNA)

            Ribonucleic acid (RNA)

1.8.14

    i.   Morpholino nucleic acid

    ii. Peptide (or peptido) nucleic acid

    iii. 4'-Thionucleic acid

1.8.15 Amino (NH_2) and carboxyl (COOH)

1.8.16 Because there are three hydroxyl groups in the glycerol to be esterified.

1.8.17 C, H, O, N, and P

1.8.18 Lipids are large enough to be considered macromolecules but they are not clas-
        sified as polymers because **they are not made by connecting monomers.**

1.8.19 Amino acids

1.8.20. The condensation of two amino acids to form a peptide bond that releases
        water.

✱✱✱✱✱✱✱✱✱✱

Chapter 2

2.8.1 Because it has six carbons and an aldehyde functional group.

2.8.2 Galactose, glucose, and fructose

2.8.3 The branching points are the main difference between them.

2.8.4 Sucrose is a disaccharide because it has two sugar units.
       The sugar units in lactose are galactose and glucose.

2.8.5 Monosaccharides (1 sugar unit), disaccharides (2 sugar units), oligosaccharides
       (2–10 sugar units), and polysaccharides (more than 10 sugar units)

2.8.6

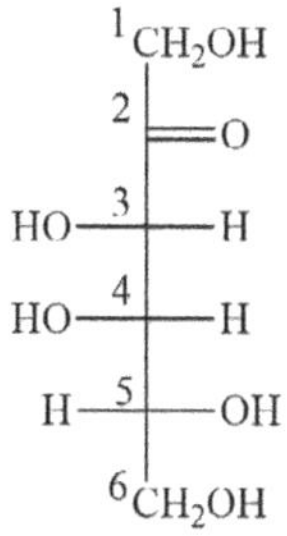

2.8.7 D-Erythrose and D-threose

2.8.8

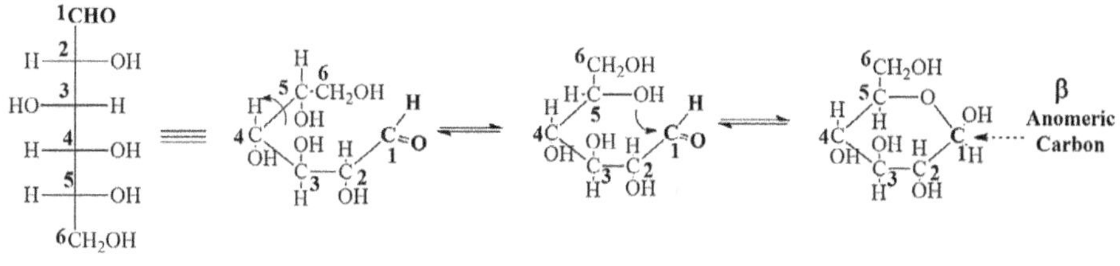

D-Glucose

2.8.9

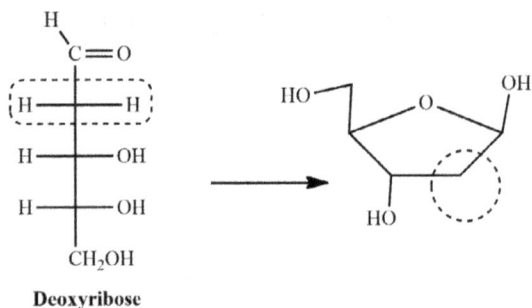

Deoxyribose

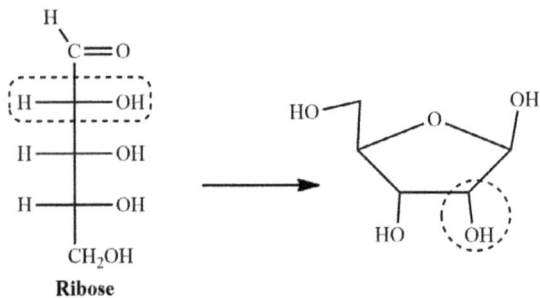

Ribose

2.8.10 Sucrose is an α(1→2) glycosidic bond.
Lactose is a β(1→4) glycosidic bond.
Maltose is an α(1→4) glycosidic bond.

2.8.11

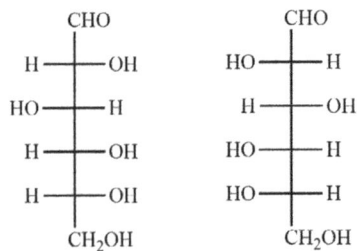

D-Glucose   L-Glucose

2.8.12

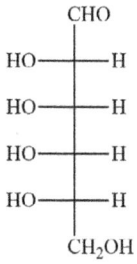

2.8.13

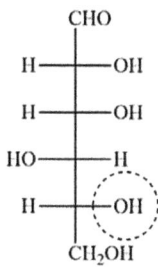 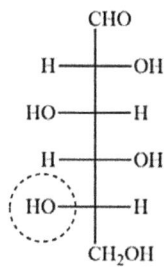 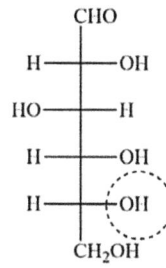

D-isomer   L-isomer   D-isomer

2.8.14

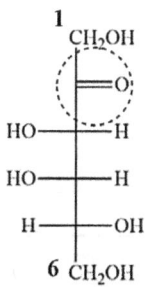 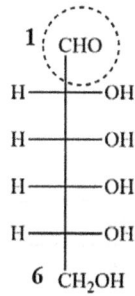

Ketohexose   Aldohexose

2.8.15

i.   2

ii. 2–10

iii. >10

2.8.16

i. It is a polysaccharide.✓

ii. It contains two different types of monosaccharides.

iii. It is a branched-chain glucose polymer.✓

2.8.17

i. Glucose and fructose

ii. Glucose

iii. Glucose

2.8.18

i. A glucose-derivative polymer

ii. A glucose polymer

iii. A glucose polymer

2.8.19

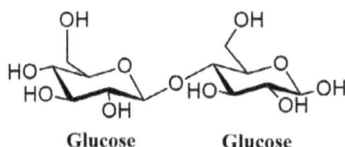

Glucose   Glucose

2.8.20 Microcrystalline cellulose (MCC)

Powdered cellulose (PC)

Low-crystallinity powdered cellulose (LCPC)

∗∗∗∗∗∗∗∗∗∗

Chapter 3

3.14.1

Column A	Column B
1. The notation for palmitic acid is ___E___	A. Long-chain carboxylic acids
2. Fatty acids are ___A___	B. Hydrogenation catalysts
3. The structures of steroids are based on ___D___	C. 20-Carbon carboxylic acids
4. Eicosanoids are a group of ___C___	D. Fused tetracyclic ring system
5. Nickel (Ni) and platinum (Pt) are ___B___	E. 16:0

3.14.2

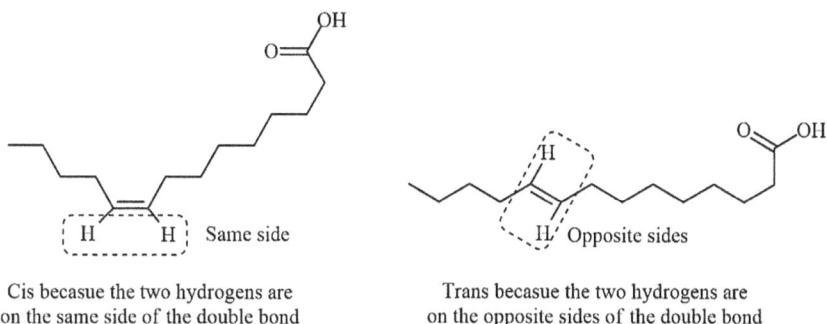

Cis becasue the two hydrogens are
on the same side of the double bond

Trans becasue the two hydrogens are
on the opposite sides of the double bond

3.14.3 16 = number of carbons in the chain

1 = number of double bonds in the structure

Δ^9 = position of the double bond between carbons 9 and 10

ω-7 = the first double bond faced when back counting is done from the last carbon in the chain, ω-carbon

3.14.4

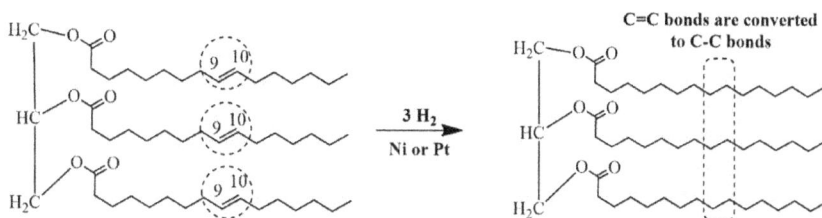

3.14.5

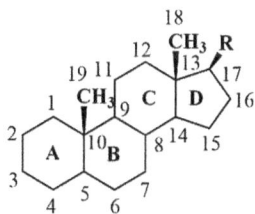

3.14.6 This fatty acid is the Ω-3 (ω-3)

3.14.7

    i.   16:1 (Δ^9) ω-7

    ii. 18:3 ($\Delta^{9,\ 12,\ 15}$) ω-3

    iii. 18:2 ($\Delta^{9,\ 12}$) ω-6

3.14.8

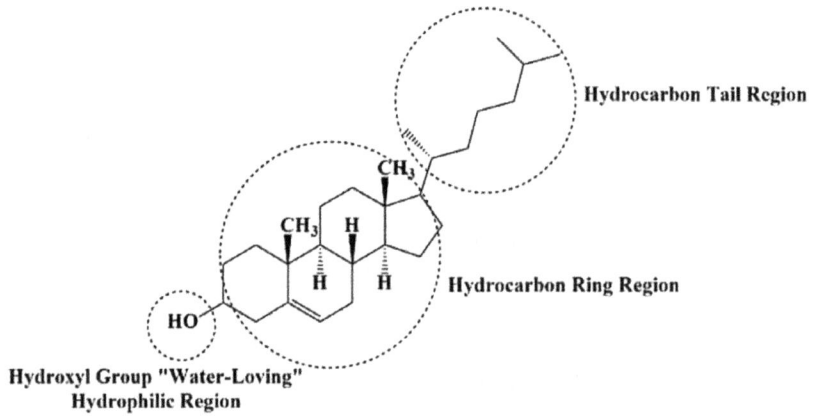

3.14.9

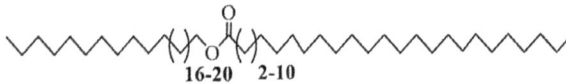

3.14.10 Waxes

3.14.11 Three fatty acids

3.14.12 Hydroxyl groups of glycerol and carboxyl groups of fatty acids

3.14.13

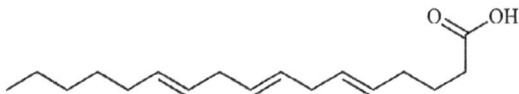

3.14.14

    i.   Oleic acid

    ii. Linolenic acid √

    iii. Arachidonic acid

3.14.15 Phosphatidic acid
3.14.16

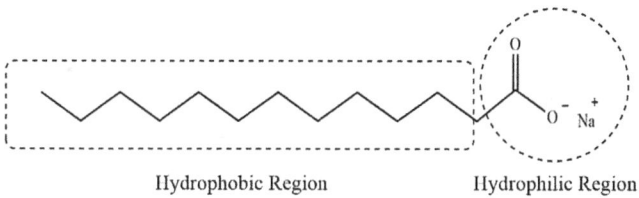

Hydrophobic Region   Hydrophilic Region

3.14.17 Vitamin A is a fat-soluble vitamin and vitamin C is a water-soluble vitamin.
3.14.18
    i.   Steroid
    ii. Lecithin√
    iii. Cephalin
3.14.19 Menthol, camphor, limonene, and pinene
3.14.20 Cholic and chenodeoxycholic acid, deoxycholic acid, and lithocholic acid

****

Chapter 4

4.11.1 Because they have a chiral carbon to which four different groups are attached. In glycine, only three different groups are attached because two out of four are hydrogens.
4.11.2

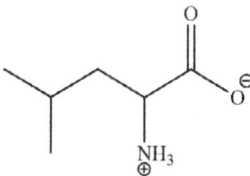

4.11.3

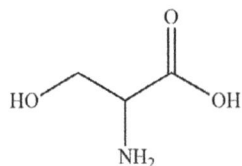

Serine

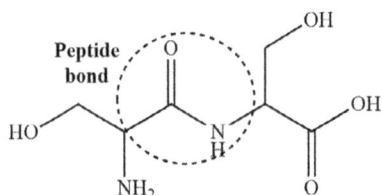

Serinylserine

4.11.4

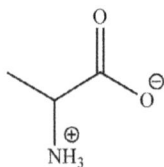 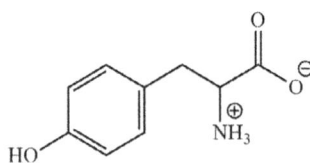

Alanine zwitterion   Tyrosine zwitterion

4.11.5

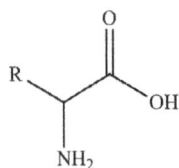

4.11.6

i.   Lysine
ii. Alanine
iii. Glycine ✓

4.11.7

i.   Proline ✓
ii. Tryptophan
iii. Histidine

4.11.8

    i.   Serine

    ii.   Phenylalanine √

    iii. Asparagine

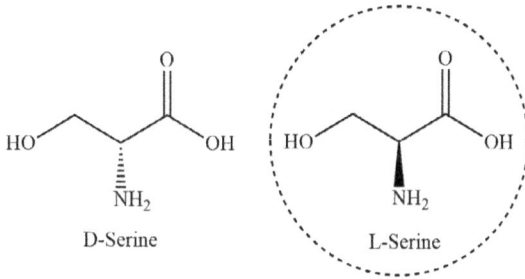

D-Serine   L-Serine

4.11.9.

4.11.10 Rs – 1 = 3 – 1 = 2.

4.11.11

    i.   Electrophoresis √

    ii.   Determination of the C-terminal amino acid

    iii. Determination of the N-terminal amino acid

4.11.12

    i.   Oxidized √

    ii.   Reduced

    iii. Hydrolyzed

4.11.13

    i.   Sanger

    ii.   Merrifield √

    iii. Strecker

4.11.14 Amino and carboxyl groups

4.11.15 DCC (dicyclohexylcarbodiimide)

4.11.16

    i.   Cellulase (catalyzes the hydrolysis of cellulose)

    ii.   Sucrase (breaks down sucrose into simple sugars)

    iii. L-Amino acid oxidase (catalyzes the stereospecific oxidative deamination of an L-amino acid substrate to an α-keto acid)

4.11.17

    i.   An enzyme √

    ii.   A hormone

    iii. A tRNA molecule

4.11.18

   i. Amino acids √

   ii. Disaccharides

   iii. Nucleotides

4.11.19

   i. Between the R group of one amino acid and the R group of the second

   ii. Between the carboxyl group of one amino acid and the amino group of the other √

   iii. Between the nitrogen atoms of the amino groups in the amino acids

4.11.20

   i. Primary structure

   ii. Secondary structure √

   iii. Tertiary structure

****

Chapter 5

5.15.1

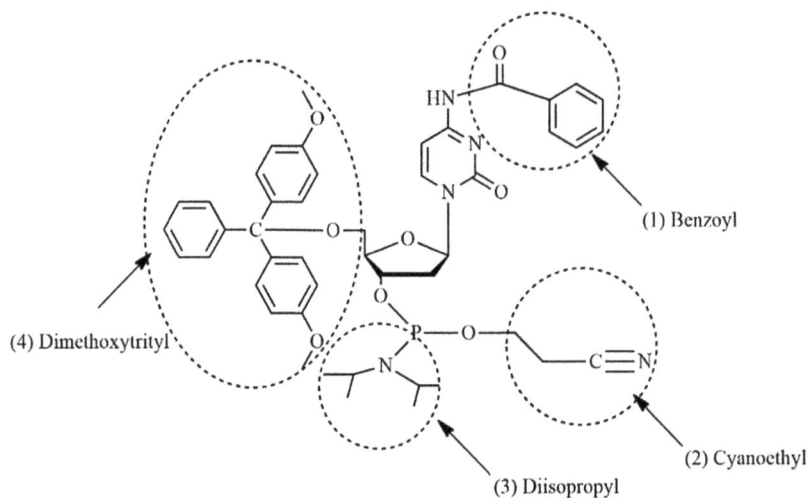

(1) Benzoyl

(4) Dimethoxytrityl

(3) Diisopropyl

(2) Cyanoethyl

5.15.2

   i. Peptide nucleic acid

   ii. Threose nucleic acid

   iii. Morpholino nucleic acid

5.15.3 G-T-G-C-G-A-T-A-C-T-A-T-A-G-C-G-G-C

5.15.4 Nucleobase, pentose sugar either ribose or deoxyribose, and phosphate

5.15.5 Three hydrogen bonds

5.15.6

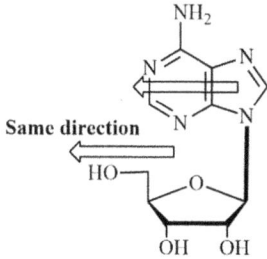
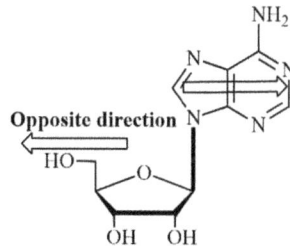

Syn-nucleoside: sugar 5'-hydroxy group and nucleobase on the same direction

Anti nucleoside: sugar 5'-hydroxy group and nucleobase on the opposite directions

5.15.7 Deblocking the 5'-DMT group and then coupling the phosphoramidite mono-mer to the first nucleoside attached to the solid support.

5.15.8

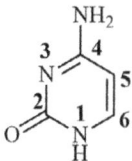
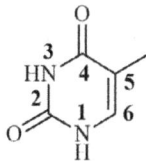
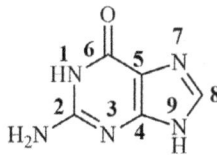

     Cytosine   **Thymine**   **Guanine**

5.15.9

DNA	RNA
Deoxyribose sugar	Ribose sugar
Thymine base	Uracil base

5.15.10 Guanine

5.15.11

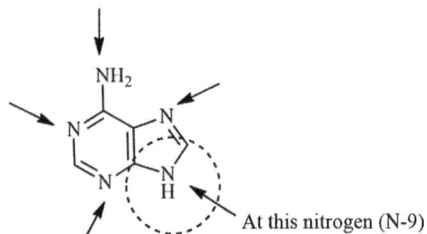

At this nitrogen (N-9)

5.15.12

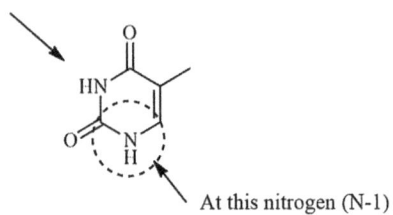

At this nitrogen (N-1)

5.15.13 Adenosine, guanosine, cytidine, and uridine

5.15.14 Cytosine and thymine

5.15.15

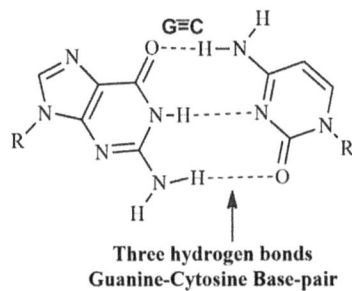

Three hydrogen bonds
Guanine-Cytosine Base-pair

5.15.16

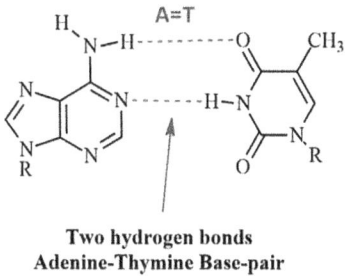

Two hydrogen bonds
Adenine-Thymine Base-pair

5.15.17

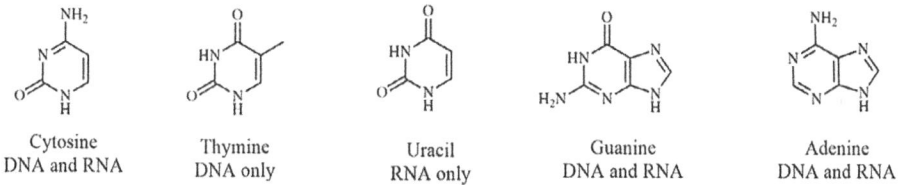

Cytosine	Thymine	Uracil	Guanine	Adenine
DNA and RNA	DNA only	RNA only	DNA and RNA	DNA and RNA

5.15.18 The amino acid sequence is Gln-Met-Pro-Trp-Pro-Leu.

5.15.19 Canada

5.15.20 7-Diethylamino-3-[4-(iodoacetamido)phenyl]-4-methylcoumarin (DCIA) and 5-(iodoacetamido)fluorescein (5-IAF)

****

Abbreviations

Ade	Adenine
ADP	Adenosine diphosphate
Ala	Alanine
Arg	Arginine
Asn	Asparagine
Asp	Aspartic acid
ASO	Antisense oligonucleotides
ATP	Adenosine triphosphate
BOC	Butyloxycarbonyl
BPA	Bisphenol A
CA	Cellulose acetate
CAP	Cellulose acetate phthalate
CMC	Carboxymethylcellulose
CN	Cellulose nitrate
CODIS	Combined DNA Index System
CPG	Controlled pore glass
CS	Cellulose sulfate
Cys	Cysteine
Cyt	Cytosine
DBS	Dibenzylidene sorbitol
DCC	Dicyclohexylcarbodiimide
DCIA	7-Diethylamino-3-[4-(iodoacetamido)phenyl]-4-methylcoumarin
DIC	Diisopropylcarbodiimide
DNA	Deoxyribonucleic acid
dNTPs	Deoxynucleotide triphosphates
EC	Ethyl cellulose
EON	Editing oligonucleotides
ESD	Electrostatic discharge
ESI	Electrospray ionization
ETYA	Eicosatetraynoic acid
FTSC	Fluorescein-5-thiosemicarbazide
α-GAL	α-Galactosidase
Gln	Glutamine
Glu	Glutamic acid
Gly	Glycine
GNA	Glycol/glycerol nucleic acids
GPLs	Glycerophospholipids
Gua	Guanine
HCN	Hydrogen cyanide
HDL	High-density lipoprotein
HEC	Hydroxyethylcellulose
His	Histidine
HPC	Hydroxypropylcellulose
HPMC	Hydroxypropylmethylcellulose
Hyp	Hydroxyproline
5-IAF	5-(Iodoacetamido)fluorescein
Ile	Isoleucine

https://doi.org/10.1515/9783111583273-007

IUPAC	International Union of Pure and Applied Chemistry
KOH	Potassium hydroxide
LCPC	Low-crystallinity powdered cellulose
LDL	Low-density lipoprotein
Leu	Leucine
LOP	L is lauric acid, O is oleic acid, and P is palmitic acid
LOS	L is linoleic acid, O is oleic acid, and S is stearic acid
LNA	Locked nucleic acid
LT	Leukotrienes
LX	Lipoxins
Lys	Lysine
MC	Methylcellulose
MCC	Microcrystalline cellulose
Met	Methionine
miRNA	MicroRNA
MNA	Morpholino nucleic acid
mRNA	Messenger ribonucleic acid
NaOH	Sodium hydroxide
NMR	Nuclear magnetic resonance
Ni	Nickel
17-ODYA	17-Octadecyonic acid
PAGE	Polyacrylamide gel electrophoresis
PC	Powdered cellulose
PCR	Polymerase chain reaction
PEG	Polyethylene glycol
PG	Prostaglandins
Phe	Phenylalanine
pI or IEP	Isoelectric point
PLs	Phospholipids
PNA	Peptide (or peptido) nucleic acid
POL	P is palmitic acid, O is oleic acid, and L is linolenic acid
Pro	Proline
Pt	Platinum
RNA	Ribonucleic acid
rRNA	Ribosomal ribonucleic acid
saRNA	Small activating RNA
Ser	Serine
siRNA	Small interfering RNA
Sn	Stereospecific numbering
SS	Solid support
SSO	Splice-switching oligonucleotide
STR	Short tandem repeat
STRs	Short tandem repeats
TGs	Triglycerides
3D	Three-dimensional or three dimensions
Thr	Threonine
Thy	Thymine
TOF	Time of flight
TNA	Threose nucleic acid

tRNA	Transfer ribonucleic acid
Trp	Tryptophan
TX	Thromboxanes
Tyr	Tyrosine
Ura	Uracil
UV-B	Ultraviolet B radiation
Val	Valine
XNAs	Xeno nucleic acids
ZN	Ziegler-Natta

Resources and Further Readings

Books

[1]   Elzagheid, M. Macromolecular Chemistry: Natural & Synthetic Polymers, Walter de Gruyter GmbH & Co KG, 2021, ISBN: 9783110762754, 2021.
[2]   Stoker, H. S. General, Organic, and Biological Chemistry, Brooks Cole, 2010, ISBN: 9780618606061.
[3]   McMurry, J., Castellion, M. E., Ballantine, D. S., Hoeger, C. A. and Virginia, E. Fundamentals of General, Organic, and Biological Chemistry, Peterson, 2007, ISBN: 0136054501.
[4]   Solomons, T. W. G., Fryhle, C. B. and Snyder, S. A. Organic Chemistry, John Wiley & Sons, 2016, ISBN: 9781118875766.
[5]   Wade, L. G. Organic Chemistry, Pearson, 2014, ISBN: 9781292021652.
[6]   McMurry, J. Organic Chemistry, Brooks Cole, 2011, ISBN: 9780840054449.
[7]   Morrison, R. T. and Boyed, R. N. Organic Chemistry, Prentice Hall, 1992, ISBN: 0136436692.
[8]   Joule, J. A. and Mills, K. Heterocyclic Chemistry, John Wiley & Sons Ltd, 2010, ISBN: 9781405193658.
[9]   Atkins, R. C. and Carey, F. A. Organic Chemistry "A Brief Course", McGraw-Hill, 2002, ISBN:0072319445.
[10]   Vollhardt, P. and Schore, N. Organic Chemistry "Structure and Function", Freeman and Company, 2014, ISBN: 139871464120275.
[11]   Brown, W. H., Iverson, B. L., Anslyn, E. V. and Foote, C. S. Organic Chemistry, Wadsworth, Cengage Learning, 2014, ISBN: 139871285052816.
[12]   Brown, W. H. and Poon, T. Organic Chemistry, John Wiley & Sons, Inc, 2016, ISBN: 9781118-875803.
[13]   Solomons, T. W. G., Fryhle, C. B. and Snyder, S. A. Organic Chemistry, John Wiley & Sons, Inc, 2013, ISBN: 9781118147399.
[14]   Blackburn, G. M., Gait, M. J., Loakes, D. and Williams, D. M. Nucleic Acids in Chemistry and Biology, RSC Publishing, 2006, ISBN: 9780854046546.
[15]   Lönnberg, H. Chemistry of Nucleic Acids, De Gruyter, 2020, ISBN: 9783110609271.

Internet Resources

– Reactions of Monosaccharides. https://chem.libretexts.org/Bookshelves/OrganicChemistry/Organic_Chemistry_(Morsch_et_al.)/25%3A_Biomolecules-_Carbohydrates/25.06%3A_Reactions_of_Monosaccharides. Accessed on June 10, 2024.
– Triglycerides structure. https://www.lipidcenter.com/pdf/Triglyceride_Structure.pdf. Accessed on June 11, 2024.
– Fatty acids, triglyceride structure, and lipid metabolism. https://www.sciencedirect.com/science/article/abs/pii/0955286395000304. Accessed on June 11,2024.
– Lipid nomenclature. https://www.lipidmaps.org/data/classification/lipid_cns.html#N. Accessed on June 11, 2024.
– Triple-Bonded Unsaturated Fatty Acids. https://page-one.springer.com/pdf/preview/10.1007/s11745-001-0740-6. Accessed on June 11, 2024.
– Bacterial wax synthesis. https://www.sciencedirect.com/science/article/abs/pii/S0734975020301828?via%3Dihub. Accessed June 11, 2024.
– Phosphatidylinositol and Related Phosphoinositides. https://lipidmaps.org/resources/lipidweb/lipidweb_html/lipids/complex/pi/index.htm. Accessed on June 12, 2024.
– Steroids. https://www.britannica.com/science/steroid. Accessed on June 12, 2024.

https://doi.org/10.1515/9783111583273-008

– Nucleoside chemistry. https://baranlab.org//images/grpmtgpdf/OHara_Jun_12.pdf. Accessed on June 12, 2024.
– Bile Acids. https://www.ncbi.nlm.nih.gov/books/NBK548626/#:~:text=The%20primary%20bile%20acids%20synthesized,deoxycholic%20acid%20and%20lithocholic%20acid. Accessed on June 13, 2024.
– Biological macromolecules review. https://www.khanacademy.org/science/ap-biology/chemistry-of-life/properties-structure-and-function-of-biological-macromolecules/a/hs-biological-macromolecules-review. Accessed on June 14, 2024.
– The Major Macromolecules. https://wou.edu/chemistry/chapter-11-introduction-major-macromolecules/. Accessed on June 14, 2024.
– Keratin. https://en.wikipedia.org/wiki/Keratin. Accessed on June 16, 2024.
– Structure of Keratin. https://link.springer.com/protocol/10.1007/978-1-0716-1574-4_5. Accessed on June 16, 2024.
– Wool, a natural biopolymer: extraction and structure–property relationships. https://www.sciencedirect.com/science/article/abs/pii/B9780323998536000164. Accessed on June 16, 2024.
– Keratin Associations with Synthetic, Biosynthetic and Natural Polymers: An Extensive Review. https://www.mdpi.com/2073-4360/12/1/32. Accessed on June 16, 2024.
– Deep Analysis of Chemical Composition of Wool with Polymeric Structure of Wool / Easy way. https://textiletrainer.com/deep-analysis-of-chemical-composition-of-wool/. Accessed on June 17, 2024.
– Collagen Structure and Stability. https://www.ncbi.nlm.nih.gov/pmc/articles/PMC2846778/. Accessed on June 17, 2024.
– Silk. https://www.swicofil.com/commerce/products/silk/276/properties. Accessed on June 17, 2024.
– Gelatin. https://water.lsbu.ac.uk/water/gelatin.html. Accessed on June 17, 2024.
– Induced Fit Model. https://www.sciencedirect.com/topics/biochemistry-genetics-and-molecular-biology/induced-fit-model. Accessed on June 17, 2024.
– Structure and Properties of Gelatin. https://encyclopedia.pub/entry/44735. Accessed on June 17, 2024.
– What are the models of enzyme action? https://www.aatbio.com/resources/faq-frequently-asked-questions/what-are-the-models-of-enzyme-action. Accessed on June 17, 2024.
– DNA Structure and Functions. https://www.sciencefacts.net/dna-structure-and-functions.html. Accessed on June 17, 2024.
– Maitotoxin. https://www.sciencedirect.com/topics/pharmacology-toxicology-and-pharmaceutical-science/maitotoxin. Accessed on July 3, 2024.
– Nucleosides and Nucleoside Analogues as Emerging Antiviral Drugs. https://www.eurekaselect.com/article/109379. Accessed on July 5, 2024.
– Monosaccharides. https://www.sciencedirect.com/topics/agricultural-and-biological-sciences/monosaccharide#:~:text=Monosaccharides%20are%20the%20simplest%20carbohydrates,%2C%20that%20is%2C%20in%20glycolysis. Accessed on July 5, 2024.
– Nucleoside anticancer drugs: the role of nucleoside transporters in resistance to cancer chemotherapy. https://pubmed.ncbi.nlm.nih.gov/14576856/. Accessed on July 6, 2024.
– Recent Advances in Nucleosides: Chemistry and Chemotherapy. https://shop.elsevier.com/books/recent-advances-in-nucleosides-chemistry-and-chemotherapy/chu/978-0-444-50951-2. Accessed on July 6, 2024.
– AIDS-driven nucleoside chemistry. https://pubs.acs.org/doi/10.1021/cr00016a004. Accessed on July 6, 2024.
– Palladium-Assisted Routes to Nucleosides. https://pubs.acs.org/doi/10.1021/cr010374q. Accessed on July 6, 2024.
– New Trends in Nucleoside Biotechnology. https://www.ncbi.nlm.nih.gov/pmc/articles/PMC3347554/. Accessed on July 6, 2024.
– Antiviral therapy. https://pubmed.ncbi.nlm.nih.gov/2890817/. Accessed on July 6, 2024.

– Wax in Biochemistry: Properties, Functions, and Contrasts with Triglycerides. https://www.creative-proteomics.com/resource/waxes-biochemistry-functions-contrasts.htm. Accessed on July 6, 2024.
– Bacterial wax synthesis. https://www.sciencedirect.com/science/article/abs/pii/S0734975020301828. Accessed on July 6, 2024.
– What Is Wedge and Dash Projection in Chemistry? https://www.thoughtco.com/wedge-and-dash-projection-definition-602137. Accessed on July 6, 2024.
– Wedge and Dash Representation. https://www.chemistrysteps.com/wedge-and-dash-representation/. Accessed on July 6, 2024.
– Depicting Carbohydrate Stereochemistry – Fischer Projections. https://chem.libretexts.org/Bookshelves/Organic_Chemistry/Organic_Chemistry_(Morsch_et_al.)/25%3A_Biomolecules-_Carbohydrates/25.02%3A_Depicting_Carbohydrate_Stereochemistry_-_Fischer_Projections. Accessed on July 6, 2024.
– Fischer and Haworth projections. https://chem.libretexts.org/Bookshelves/Organic_Chemistry/Book%3A_Organic_Chemistry_with_a_Biological_Emphasis_v2.0_(Soderberg)/03%3A_Conformations_and_Stereochemistry/3.09%3A_Fischer_and_Haworth_projections. Accessed on July 6, 2024.
– Fischer, C., Haworth, and Chair forms of Carbohydrates. https://www.chemistrysteps.com/converting-fischer-haworth-chair-carbohydrates/. Accessed on July 6, 2024.
– Keto- and Enol Tautomerism in Sugars. https://www.news-medical.net/life-sciences/Keto-and-Enol-Tautomerism-in-Sugars.aspx. Accessed on July 7, 2024.
– Wohl Degradation. https://www.chemistrysteps.com/wohl-degradation/. Accessed on July 7, 2024.
– Isomaltose – A definition and examples. https://www.sugarnutritionresource.org/news-articles/isomaltose-a-definition-and-examples. Accessed on July 7, 2024.
– Trehalose. https://www.sciencedirect.com/topics/chemistry/trehalose. Accessed on July 7, 2024.
– What Is Lactose? https://foodinsight.org/what-is-lactose/. Accessed on July 7, 2024.
– What Are Oligosaccharides? All You Need to Know. https://www.healthline.com/nutrition/oligosaccharides#what-they-are. Accessed on July 8, 2024.
– Maltodextrin. https://www.acs.org/molecule-of-the-week/archive/m/maltodextrin.html. Accessed on July 8, 2024.
– Maltotetraose. https://www.caymanchem.com/product/24975. Accessed on July 8, 2024.
– Starch. https://www.britannica.com/science/starch. Accessed on July 8, 2024.
– What Is Glycogen? https://www.webmd.com/a-to-z-guides/what-is-glycogen. Accessed on July 8, 2024.
– Pullulan. https://www.fao.org/fileadmin/templates/agns/pdf/jecfa/cta/65/pullulan.pdf. Accessed on July 8, 2024.
– What Is Lignin? https://www.lignoworks.ca/content/what-lignin/. Accessed on July 8, 2024.
– Peptide Bioconjugation Protocols and Techniques – A Simple Guide. https://blog.papyrusbio.com/peptide-bioconjugation/. Accessed on July 9, 2024.
– Synthesis of multifunctional lipid–polymer conjugates: application to the elaboration of bright far-red fluorescent lipid probes. https://pubs.rsc.org/en/content/articlehtml/2014/ra/c4ra01334d. Accessed on July 9, 2024.
– Evaluation of the effects of chemically different linkers on hepatic accumulations, cell tropism and gene silencing ability of cholesterol-conjugated antisense oligonucleotides. https://www.sciencedirect.com/science/article/abs/pii/S0168365916300608. Accessed on July 9, 2024.
– Gamma-linolenic acid. https://www.tuscany-diet.net/lipids/fatty-acids/gamma-linolenic-acid/. Accessed on July 10, 2024.
– Fatty Acids. https://chem.libretexts.org/Courses/American_River_College/CHEM_309%3A_Applied_Chemistry_for_the_Health_Sciences/11%3A_Lipids_-_An_Introduction/11.01%3A_Fatty_Acids. Accessed on July 11, 2024.
– Wax. https://chem.libretexts.org/Bookshelves/Biological_Chemistry/Supplemental_Modules_%28Biological_Chemistry%29/Lipids/Non-glyceride_Lipids/Wax. Accessed on July 12, 2024.

– Carnauba wax. https://atamankimya.com/sayfalaralfabe.asp?LanguageID=2&cid=3&id=2868&id2=3610. Accessed on July 12, 2024.
– Triglycerides. https://chem.libretexts.org/Bookshelves/Introductory_Chemistry/Introductory_Chemistry_(CK-12)/26%3A_Biochemistry/26.08%3A_Triglycerides. Accessed on July 12, 2024.
– Naming triglycerides. https://foodsciencetoolbox.com/naming-triglyceridies/. Accessed on July 12, 2024.
– Nomenclature of Lipids. https://iupac.qmul.ac.uk/lipid/lip1n2.html#p11. Accessed on July 12, 2024.
– Phosphatidylinositol. https://en.wikipedia.org/wiki/Phosphatidylinositol. Accessed on July 13, 2024.
– Phosphoglycerides or Phospholipids. https://chem.libretexts.org/Bookshelves/Biological_Chemistry/Supplemental_Modules_(Biological_Chemistry)/Lipids/Glycerides/Phosphoglycerides_or_Phospholipids. Accessed on July 13, 2024.
– Phosphatidylserine. https://en.wikipedia.org/wiki/Phosphatidylserine. Accessed on July 13, 2024.
– Phosphoinositide 3-kinase signaling in the vertebrate retina. https://www.sciencedirect.com/science/article/pii/S0022227520313729. Accessed on July 13, 2024.
– Lipoid. https://lipoid.com/en/products/synthetic-phospholipids/phosphatidic-acids/. Accessed on July 13, 2024.
– Sphingolipid. https://www.britannica.com/science/sphingolipid. Accessed on July 13, 2024.
– Sphingolipids. https://phys.libretexts.org/Courses/University_of_California_Davis/Biophysics_241%3A_Membrane_Biology/01%3A_Lipids/1.05%3A_Sphingolipids. Accessed on July 13, 2024.
– Eicosanoid. https://en.wikipedia.org/wiki/Eicosanoid. Accessed on July 14, 2024.
– Thromboxane. https://www.sciencedirect.com/topics/medicine-and-dentistry/thromboxane. Accessed on July 15, 2024.
– Leukotrienes: Structure, Functions, and Modulation Strategies. https://www.creative-proteomics.com/blog/index.php/leukotrienes-structure-functions-and-modulation-strategies/. Accessed on July 15, 2024.
– Prostaglandins: Structure, Functions, and Analytical Methods. https://www.creative-proteomics.com/resource/prostaglandins-structure-functions-analytical-methods.htm. Accessed on July 15, 2024.
– Lipoxin. https://en.wikipedia.org/wiki/Lipoxin. Accessed on July 15, 2024.
– Steroid numbering system and nomenclature. https://www.britannica.com/science/steroid/Steroid-numbering-system-and-nomenclature. Accessed on July 15, 2024.
– Terpenes. https://www.chem.ucalgary.ca/courses/351/Carey5th/Ch26/ch26-4-1.html. Accessed on July 15, 2024.
– Terpenoid. https://www.sciencedirect.com/topics/neuroscience/terpenoid. Accessed on July 16, 2024.
– What are vitamins, and how do they work? https://www.medicalnewstoday.com/articles/195878. Accessed on July 16, 2024.
– Vitamin, A. https://courses.lumenlearning.com/suny-nutrition/chapter/12-6-vitamin-a/. Accessed on July 16, 2024.
– Biomolecules: Protein 1. https://www2.chem.wisc.edu/deptfiles/genchem/netorial/modules/biomolecules/modules/protein1/prot13.htm. Accessed on July 18, 2024.
– Alpha-Helix and Beta-Sheet. https://byjus.com/chemistry/alpha-helix-and-beta-sheet/. Accessed on July 20, 2024.
– Wool structure and morphology. https://www.sciencedirect.com/science/article/abs/pii/B9780128240564000133. Accessed on July 20, 2024.
– Notes on Collagen. https://earthwormexpress.com/2020/09/27/notes-on-collagen/. Accessed on July 20, 2024.
– Current Insights into Collagen Type I. https://www.mdpi.com/2073-4360/13/16/2642. Accessed on July 20, 2024.
– What Are the Different Types of Collagen and Their Benefits? https://ca.doseandco.com/blogs/science/what-are-the-different-types-of-collagen-and-their-benefits. Accessed on July 20, 2024.
– Structure and Properties of Gelatin. https://encyclopedia.pub/entry/44735. Accessed on July 21, 2024.

– Nucleic acid. https://www.britannica.com/science/nucleic-acid/Deoxyribonucleic-acid-DNA. Accessed on July 23, 2024.
– Discovery of DNA Structure and Function: Watson and Crick. https://www.nature.com/scitable/topic page/discovery-of-dna-structure-and-function-watson-397/. Accessed on July 23, 2024.
– Reagents for nucleic acids synthesis. https://labchem-wako.fujifilm.com/us/catalog/pdf/catalog_0026. pdf. Accessed on July 23, 2024.
– Sugar pucker correlates with phosphorus-base distance. https://x3dna.org/highlights/sugar-pucker-correlates-with-phosphorus-base-distance. Accessed on July 24, 2024.
– Nucleic Acid Drugs – Current Status, Issues, and Expectations for Exosomes. https://www.ncbi.nlm.nih. gov/pmc/articles/PMC8508492/. Accessed on July 24, 2024.
– Modified nucleic acids: replication, evolution, and next-generation therapeutics. https://bmcbiol.biomed central.com/articles/10.1186/s12915-020-00803-6. Accessed on July 24, 2024.
– Locked Nucleic Acid. https://www.syngenis.com/locked-nucleic-acid/. Accessed on July 24, 2024.
– Antisense Oligonucleotides: An Emerging Area in Drug Discovery and Development. https://www.ncbi. nlm.nih.gov/pmc/articles/PMC7355792/. Accessed on July 24, 2024.
– End-Labeling Oligonucleotides with Chemical Tags After Synthesis. https://www.ncbi.nlm.nih.gov/pmc/ articles/PMC5026237/. Accessed on July 24, 2024.
– Radioactive-labeling-nucleic-acids-autoradiography.https://istudy.pk/radioactive-labeling-nucleic-acids-autoradiography/. Accessed on July 24, 2024.
– Techniques for Oligonucleotide Analysis. https://www.technologynetworks.com/genomics/articles/tech niques-for-oligonucleotide-analysis-342060. Accessed on July 24, 2024.
– DNA Replication. https://socratic.org/biology/dna-structure-and-function/dna-replication. Accessed on July 25, 2024.
– Your genetic code has lots of "words" for the same thing – information theory may help explain the redundancies. https://theconversation.com/your-genetic-code-has-lots-of-words-for-the-same-thing-information-theory-may-help-explain-the-redundancies-209471. Accessed on July 25, 2024.
– Genetic code. https://byjus.com/biology/genetic-code/. Accessed on July 25, 2024.
– The genetic code. https://www.khanacademy.org/science/ap-biology/gene-expression-and-regulation /translation/a/the-genetic-code-discovery-and-properties. Accessed on July 25, 2024.
– What is DNA fingerprinting? https://www.civilsdaily.com/news/dna-fingerprinting/. Accessed on July 25, 2024.
– DNA fingerprinting. https://www.britannica.com/science/DNA-fingerprinting. Accessed on July 25, 2024.
– DNAs Secret Code. https://www.seminarsonly.com/Engineering-Projects/Chemistry/dna-secret-code. php. Accessed on July 25, 2024.
– Nucleic Acids. https://bio.libretexts.org/Bookshelves/Biotechnology/Bio-OER_(CUNY)/02%3A_Chemistry/ 2.10%3A_Nucleic_Acids. Accessed on July 27, 2024.
– Biochem. Lab-Lipids. https://quizlet.com/ph/847589906/biochem-lab-lipids-flash-cards/. Accessed on July 27, 2024.
– Color Presenting Products of Amino Acids Reactions- Qualitative Tests. http://commaterial.org/article/10. 11648/j.mc.20180604.13. Accessed on July 27, 2024.
– Laboratory Activities to Introduce Carbohydrates Qualitative Analysis to College Students. https://www. semanticscholar.org/paper/Laboratory-Activities-to-Introduce-Carbohydrates-to-Elzagheid/ec16920c c294ed5a22396e9f0ee6cafd0c81d0ab?p2df. Accessed on July 27, 2024.

Appendices

Appendix A

Qualitative Tests for Biomacromolecules and Their Monomers

Test name	Biomacromolecule type	Positive result color
Biuret test A colorimetric response in which the color changes from blue to purple or violet. In an alkaline environment, the biuret reagent's cupric (Cu^{2+}) ions bind to nitrogen atoms in protein-peptide linkages, forming a violet-colored copper coordination complex. The intensity of the resulting purple color depends on the concentration of peptide bonds in the solution.	Proteins or peptides	Purple color
Dische's diphenylamine test Diphenylamine is a colorless (clear) reagent used to detect the presence of nucleic acid. It is necessary to heat the sample test tubes. This helps to weaken the nucleic acid bindings, allowing the reagent to intercalate with them.	Nucleic acids	DNA (dark blue) RNA (green color)
Huble's iodine test Unsaturated fatty acids can add halogens (iodine) to the double bonds to form halogenated derivatives, while saturated fatty acids cannot. Oil contains a higher percentage of unsaturated fatty acids than solid fat, so oils can react with more iodine.	Unsaturated fatty acids in oils	Semi-brown color disappears with the addition of unsaturated fatty acids
Iodine test Iodine reagent (IKI) distinguishes starch and glycogen from other monosaccharides and polysaccharides. The reagent yields a blue-black color in the presence of starch. Glycogen reacts with the reagent to give a brown-black color. Other polysaccharides and monosaccharides yield a yellow-orange color. Amylose, in starch, is responsible for the reaction with iodine. Its helices bind iodine atoms in the solution and produce an amylose-iodine complex.	Starch	Blue-black color

https://doi.org/10.1515/9783111583273-009

(continued)

Test name	Biomacromolecule type	Positive result color
Lead sulfide test A biochemical test that detects amino acids such as cysteine and cystine. The test is a specific test for detecting the S–S group in cystine and the S–H group in cysteine.	Cysteine and cystine or protein-containing cysteine and cystine	Black precipitate
Molisch's test A broad test that detects the presence of carbohydrates. If the test results are negative, sugars in the sample are omitted. It is a useful test for determining whether compounds can be dehydrated to furfural or hydroxymethylfurfural in the presence of H_2SO_4. Alpha-naphthol interacts with the cyclic aldehyde to produce purple-colored condensation products.	Carbohydrates or sugars	A purple color appears at the interface between the sugar and acid
Ninhydrin test This determines the presence of amino acids in unknown samples. This assay is also used in solid-phase peptide synthesis to assess protein protection during amino acid analysis.	Amino acids	Blue-purple color
Salkowski's test and Liebermann-Burchard test In the presence of strong sulfuric acid and acetic anhydride, cholesterol is dehydrated, resulting in a colored product.	Cholesterol	Blue-green color
Saponification test Lipid triglycerides can react with the alkali NaOH to create soap and glycerol in the presence of ethanol. This reaction is sometimes called alkaline hydrolysis of esters.	Lipids	Froth appearance
Seliwanoff's test This is used to distinguish between ketoses and aldoses. For example, fructose (ketose) produces a cherry red color, whereas glucose (aldose) produces a negative result with no cherry red. However, if heated for more than 5 min, sucrose (a combination of fructose and glucose) produces a cherry red color as well.	Ketohexoses	Deep cherry color

Appendix B

List of Selected Reagents Used in Biomacromolecule Synthesis

Abbreviation or chemical abbreviation	Full name	Use
Ac_2O in THF	10% Acetic anhydride/tetrahydrofuran solution	Capping reagent
Acridine orange stain	3,6-Bis(dimethylamino)acridine	Fluorescent nucleic acid binding dye
Beaucage reagent	3H-1,2-Benzodithiol-3-one1,1-dioxide	Sulfurizing reagent
BOP	Benzotriazol-1-yloxytris(dimethylamino) phosphonium hexafluorophosphate	Peptide synthesis
BTT	5-Benzylthio-1H-tetrazole	Coupling reaction activator especially in oligonucleotide synthesis
CH_3NH_2	40% Methylamine solution	Cleavage/deprotection reagent
CMPI (the Mukaiyama condensation reagent)	2-Chloro-1-methylpyridinium iodide	Peptide coupling
Congo red (amyloidophylic dye)	Sodium salt of benzidinediazo-bis-1-naphthylamine-4-sulfonic acid	Stains β sheet aggregates
DCA in $C_6H_5CH_3$	Dichloroacetic acid:toluene (3:97)	Deblocking solution
DCC	Dicyclohexylcarbodiimide	A coupling reagent used in the formation of amide bonds
DIPEA	N-Ethyldiisopropylamine	Base reagent
DMTrCl	4,4'-Dimethoxytrityl chloride	Tritylation reagent.
HATU	Hexafluorophosphate azabenzotriazole tetramethyl uronium	Utilized to create an active ester from a carboxylic acid in peptide coupling chemistry
I_2 in C_5H_5N/H_2O	Iodine in pyridine-water	Oxidation reagent
Millon's reagent	Mercuric nitrate ($Hg[NO_3]_2$)	Detect the presence of soluble proteins
MMTrCl	4-Methoxytrityl chloride	Tritylation reagent
NH_4OH	25% Ammonia solution	Cleavage/deprotection reagent

(continued)

Abbreviation or chemical abbreviation	Full name	Use
PADS	Bis(phenylacetyl) disulfide	Sulfurizing reagent
PI	Propidium iodine or 3,8-diamino-5-[3-(diethylmethylammonio)propyl]-6-phenylphenanthridinium diiodide	Fluorescent intercalating agent
PPTS	Pyridinium p-toluenesulfonate	Acidic catalyst for THP protection
Sudan III	1-[(4-Phenyldiazenylphenyl)-diazenyl] naphthalen-2-ol	A fat-soluble dye used for demonstrating triglycerides in frozen sections
Sudan Black B	(2,2-Dimethyl-1,3-dihydroperimidin-6-yl)-(4-phenylazo-1-naphthyl)diazene	Stains phospholipids and intracellular lipids
TBAF	Tetrabutylammonium fluoride	Utilized to get rid of protective groups for silyl ether It also serves as a moderate base and a phase transfer catalyst
TCA or DCA	Dichloroacetic acid or trichloroacetic acid	Detritylation reagent
TIPDSiCl$_2$	1,3-Dichloro-1,1,3,3-tetraisopropyldisiloxane	Silylation reagents

Index

https://doi.org/10.1515/9783111583273-010

www.ingramcontent.com/pod-product-compliance
Lightning Source LLC
Chambersburg PA
CBHW081531220326
41598CB00036B/6397